INFECTIOUS DISEASES AND MICROBIOLOGY

COVID-19

INFECTIOUS DISEASES AND MICROBIOLOGY

Additional books and e-books in this series can be found on Nova's website under the Series tab.

INFECTIOUS DISEASES AND MICROBIOLOGY

COVID-19

SUJITH OVALLATH

Copyright © 2020 by Nova Science Publishers, Inc.

All rights reserved. No part of this book may be reproduced, stored in a retrieval system or transmitted in any form or by any means: electronic, electrostatic, magnetic, tape, mechanical photocopying, recording or otherwise without the written permission of the Publisher.

We have partnered with Copyright Clearance Center to make it easy for you to obtain permissions to reuse content from this publication. Simply navigate to this publication's page on Nova's website and locate the "Get Permission" button below the title description. This button is linked directly to the title's permission page on copyright.com. Alternatively, you can visit copyright.com and search by title, ISBN, or ISSN.

For further questions about using the service on copyright.com, please contact:
Copyright Clearance Center
Phone: +1-(978) 750-8400   Fax: +1-(978) 750-4470   E-mail: info@copyright.com.

NOTICE TO THE READER

The Publisher has taken reasonable care in the preparation of this book, but makes no expressed or implied warranty of any kind and assumes no responsibility for any errors or omissions. No liability is assumed for incidental or consequential damages in connection with or arising out of information contained in this book. The Publisher shall not be liable for any special, consequential, or exemplary damages resulting, in whole or in part, from the readers' use of, or reliance upon, this material. Any parts of this book based on government reports are so indicated and copyright is claimed for those parts to the extent applicable to compilations of such works.

Independent verification should be sought for any data, advice or recommendations contained in this book. In addition, no responsibility is assumed by the Publisher for any injury and/or damage to persons or property arising from any methods, products, instructions, ideas or otherwise contained in this publication.

This publication is designed to provide accurate and authoritative information with regard to the subject matter covered herein. It is sold with the clear understanding that the Publisher is not engaged in rendering legal or any other professional services. If legal or any other expert assistance is required, the services of a competent person should be sought. FROM A DECLARATION OF PARTICIPANTS JOINTLY ADOPTED BY A COMMITTEE OF THE AMERICAN BAR ASSOCIATION AND A COMMITTEE OF PUBLISHERS.

Additional color graphics may be available in the e-book version of this book.

Library of Congress Cataloging-in-Publication Data

ISBN: 978-1-53618-691-8
Names: Ovallath, Sujith, author.
Title: Covid-19 / Sujith Ovallath.
Description: New York : Nova Science Publishers, [2020] | Series:
    Infectious diseases and microbiology | Includes bibliographical
    references and index. |
Identifiers: LCCN 2020041693 (print) | LCCN 2020041694 (ebook) | ISBN
    9781536186918 (paperback) | ISBN 9781536187298 (adobe pdf)
Subjects: LCSH: COVID-19 (Disease) | COVID-19 (Disease)--Epidemiology. |
    COVID-19 (Disease)--Psychological aspects. | COVID-19
    (Disease)--Economic aspects. | Epidemics--China.
Classification: LCC RA644.C67 O93 2020 (print) | LCC RA644.C67 (ebook) |
    DDC 616.2/414--dc23
LC record available at https://lccn.loc.gov/2020041693
LC ebook record available at https://lccn.loc.gov/2020041694

Published by Nova Science Publishers, Inc. † New York

CONTENTS

Preface		vii
About the Author		ix
Chapter 1	Introduction	1
Chapter 2	Epidemiology of COVID-19	11
Chapter 3	Clinical Presentation of COVID-19	21
Chapter 4	The Virus	27
Chapter 5	Pathogenesis and Pathology	33
Chapter 6	Diagnosis of COVID-19	41
Chapter 7	Transmission and Prevention of Transmission of COVID-19	47
Chapter 8	Management of COVID-19	57
Chapter 9	Prognosis of COVID-19	85
Chapter 10	Kerala Health Care System and COVID-19	89
Chapter 11	Health Care Professionals and COVID-19	95
Chapter 12	Psychological Impact of COVID-19	99

Chapter 13	Socio - Economic Impact of COVID-19	**103**
Chapter 14	Future Directions	**107**
Index		**111**

PREFACE

Novel corona virus infection was started in Wuhan, china in late 2019 and spread globally. This pandemic has resulted in shutdown of almost one fifth of global population movement as a measure to prevent the spread but, despite of the best efforts infection continues to spread. Even the countries with the best health infrastructure are affected, resulting in huge economic burden to the society.

In February 2020, the world Health organization (WHO) announced that COVID-19 is the official name of the disease and explained that *CO* stands for *corona*, *VI* for *virus* and *D* for *disease*, while *19* is for the year that the outbreak was first identified; that is 30 December 2019. While the disease is named COVID-19, the virus that causes it is named severe acute respiratory syndrome corona virus 2 or SARS-CoV-2. The virus was initially referred to as the 2019 novel coronavirus or 2019-nCoV. The WHO additionally uses "the COVID-19 virus" and "the virus responsible for COVID-19" in all its communications.

It is interesting to say that this book is being written during the strict lock down period in India as announced by the government of India, (for the next 21 days which started 4 days back) in order to prevent the spread of the pandemic in the country. Our hospital has been converted to a full time corona care centre as the number of cases of corona are increasing in this part of India. As of today a total of 621090 people are affected in the world with COVID-19 with a reported mortality of 28662 with maximum death being reported in USA followed by Italy, Spain and China. India has a total of 933 cases reported as of today with a death rate of 20. Hope we will succeed in this war against the Viruses as the existence of human race will depend on the success stories in coming months. As the scientific literature about Covid19 is rapidly evolving, some of the information given in the book may not be up-to-date by the time of publication, in view of the extensive research undergoing in different parts of the world.

Keywords: COVID-19, Novel corona virus, Wuhan epidemic, acute respiratory distress syndrome.

27-03-2020
Prof. Sujith Ovallath
Kerala, India

ABOUT THE AUTHOR

Prof. Sujith Ovallath, MBBS, MD (MEDICINE), DNB (NEUROLOGY), DM (NEUROLOGY), Clinical Fellowship in Movement Disorders, University of British Columbia.

Currently Working as Professor of Medicine and Head of Neurology, Kannur Medical College, India.

He has Passed MBBS in 1993 from Calicut Medical College, with Gold Medal in Medicine. He has completed MD in Medicine, DNB Neurology and DM neurology from same institute. Dr Sujith has completed a two year clinical fellowship in Movement Disorders from the University of British Columbia, Canada. He serves in editorial board of many journals and authored several international papers, and book chapters.

Chapter 1

INTRODUCTION

In early December2019 a new type of pneumonia like illness was reported from Wuhan city, the capital of Hubei, province in China. Within a span of two weeks five such cases were admitted and an outbreak was suspected. Out of these five cases, one patient died, and all of them had features of acute respiratory distress. Within next two weeks 41 patients admitted in the hospital, who were admitted for other reasons, also developed similar flu like respiratory symptoms and were thought to have infected with a virus [1]. A nosocomial transmission was suspected and the etiological agent was sought for. On December 31, an alert was issued by the Wuhan Municipal Health Commission. A rapid response team was advised to study the Wuhan epidemic by the Chinese Center for Disease Control and Prevention (China CDC). World Health organization (WHO) was also notified.

The early cases were thought to be linked to the Huanan Seafood and wet animal Market, and the market was shut down on January 1, 2020 [2]. On January 7, 2020, the causative pathogen was identified as a novel corona virus and was named as "severe acute respiratory syndrome corona virus 2 (SARS-CoV-2)." This name was chosen because the virus is genetically related to the corona virus responsible for the SARS outbreak of 2003. Though related, the two viruses are different [3]. On January 2020, China's "National Infectious Diseases Law" was amended to make 2019-novel corona virus diseases a Class B notifiable disease [3]. The outbreak was declared as a Public Health Emergency of International Concern on 30 January 2020 by WHO. WHO also announced "COVID-19" as the name of this new disease on 11 February 2020 [4].

COVID-19 rapidly spread from Wuhan to the entire country in just 30 days. The speed of spread and expanding number of cases overwhelmed health and public health services in China [5]. By mid of February, more than 50,000 cases with laboratory confirmed COVID-19 were detected in China, with a mortality exceeding 1600. It has spread to all 34 provinces in China within a span of one month. The rapid spread was attributed to the spring festival travel, in which an estimated 5 million people traveled from Wuhan to throughout the country, or outside. The disease started to spread to other countries also. The countries with direct air travel from Wuhan started to get affected first followed by other countries through inter personal transmission. Italy which had direct business relation with the fashion industry of Wuhan was worst affected. At the time of writing this chapter 198 countries in the world are affected by this pandemic. At the time of writing of this chapter the total estimated number of infected people has exceeded 600,000 and the death rate has exceeded 27000.

The virus continues to spread in spite of the travel ban and other social isolation measures, imposed in order to prevent the spread, on one fifth of the world population. WHO has declared COVID-19 as a pandemic on 11th march 2020 [6].

INFECTIVE AGENT AND MODE OF SPREAD

Corona viruses (CoVs) are, single stranded RNA viruses covered by a lipid envelope COVID-19 represents the seventh member of the corona virus family that infects humans and has been classified under the orthocoronavirinae subfamily. The CoVs genome, is relatively large in comparison to other viruses, ranging from 26 to 32 kilo bases in length [7, 8]. Epidemiologically, 2019-nCoV is highly infectious with about 3 h survive time in the air. The main mode of transmission is by droplet spread through coughs and sneeze mainly through person to person transmission. The incubation period after infection is generally 4–8 days but may range up to 14 days [9]. All age groups are susceptible to the virus, but elderly patients and patients with comorbidities are more likely to experience severe illness [9]. A large proportion of people infected with this virus remain asymptomatic after infection. The people who are primary, asymptomatic and those in the incubation period are the main sources of infection, to the public. These asymptomatic individuals are of critical important in the epidemiological preventive strategies. Apart from the respiratory droplets, the fecal-oral route of transmission is considered important especially during the convalasent period [10]. Vertical transmission from mothers and infants also has been confirmed as another possible route. Some researchers also warn that the transmission through the ocular surface is also a possibility in view of the fact that-human conjunctival epithelium can get contaminated by infectious droplets and body fluids. The rate of decay of the virus was evaluated and found no viable viruses were detected after four hours on copper, 24 hours on cardboard, 72 hours on stainless steel, and 72 hours on plastic. However, detection rates did not reach 100% and varied between surface type (limit Estimation of the rate of decay with a Bayesian regression model suggests that viruses may remain viable up to 18 hours on copper, 55 hours on cardboard, 90 hours on stainless steel, and over 100 hours on plastic. The virus remained viable in aerosols up to three hours [11].

SYMPTOMS AND SIGNS

Clinical symptoms generally start after an average of 5 days after exposure with flue like features. Common symptoms are fever, ,dry cough,myalgia,back pain, nausea with or without vomiting, abdominal discomfort, diarhoea,loss of smell, loss of taste, anorexia and fatigue[12, 13, 14, 15] less common symptoms include sore throat, runny nose and sneezing. Loss of smell and alteration in taste are common among infected patients who remain otherwise asymptomatic.

A few patients develop features of pneumonia with tachypnoea, tachycardia, dyspnoea and chest discomfort generally after five days. About 5% of infected go in for critical illness requiring ventilation with features of acute respiratory distress syndrome [16, 17]. A few will develop myocarditis, pericarditis, cardiac failure, tachyarrythmeas, renal failure, and liver involvement. A few will go in for multi-organ failure and die. Mortality rate varies from country to country (0.9 to 11%) and will depend on the efficiency of the treating team and facilities available [17, 18, 19].

DIAGNOSIS

The WHO has published definite protocols to be followed for the diagnosis of COVID-19 infection [20]. The standard method of testing is real-time revers transcription polymerase reaction [20].

The specimen is obtained by a nasopharyngeal swab. Other specimens that can be used include a nasal swab or sputum or bronchoalveolar lavage. Results can be obtained within a few hours to two days. Blood tests can be used, but these require two blood samples taken two weeks apart and the results have little immediate value. Chinese scientists were able to isolate a strain of the corona virus and publish the genetic sequence so that laboratories across the world could independently develop polymerase chain reaction A private laboratory in India has developed an antibody test kit. Several labs in the world are in the process of developing antibody test kits. CT scan often be very helpful in diagnosis of COVID-19. Lung involvement in COVID-19 manifests with chest CT imaging abnormalities, even in asymptomatic patients, with rapid evolution from focal unilateral to diffuse bilateral ground-glass opacities that progressed to or co-existed with consolidations within 1–3 weeks [21].

MANAGEMENT

Isolation of the infected patient remains the best strategy to control the person-to-person transmission of COVID-19 infection. At the time of writing this manuscript, there are no specific antiviral drugs or vaccine against COVID-19 infection though there are several experimental therapies being administered in different parts of the world. One of the main options available is using broad-spectrum antiviral drugs like Nucleoside analogues and also HIV-protease inhibitors that could attenuate virus infection until the specific antiviral becomes available [22]. In a trial involving 75 patients who were administrated existing antiviral drugs. For 3-14 days with 75 mg oseltamivir, 500 mg lopinavir, 500 mg ritonavir and the intravenous administration of 0·25 g ganciclovir showed some encouraging results [23]. Another invitro study showed that the broad-spectrum antiviral remdesivir and chloroquine, highly effective in the control of 2019-nCoV infection. The above drugs which are already in use can be tried to treat COVID-19 infection [24] EIDD-2801 a compound that has shown high therapeutic potential against seasonal and pandemic influenza virus infections represents another potential drug to be considered for the treatment of COVID-19 [25]. A combination of azithromycin and chloroquine also has been found useful in one trial [26]. Symptomatic management, treatment of secondary bacterial infection, proper fluid and electrolyte management, ventilator care, treatment of ARDS also play crucial role in critically ill cases of COVID-19. Avoidence of use of Ibuprofen and other NSAIDs are also recommended in COVID-19 cases as some of these agents were reported to be associated with more stormy course in Italian COVID positive cases. Several other agents are under trial for COVID-19 cases, but convincing results to recommend these agents are not yet to be available at the time of preparation of this manuscript.

PREVENTION

In view of the rapidly spreading pandemic, preventive measures are utmost important. Public education remains the most important intervention to be carried out to prevent spread of the infection. The importance of hand hygiene, frequent hand washing with soap and water, cough etiquette, social distancing, importance of using masks, avoiding touching the face , staying at home, isolation of positive cases as well as cases with contact history should be highlighted. Locking down of the country and isolation of the areas with infection had played a major role in china in preventing spread of this pandemic. Health care workers should take every precaution like proper use of personal protective devices, and measures to avoid patient to patient transmission. Asymptomatic carriers are the most important link in causing community spread. So social isolation need to be carried out at the government level.

REFERENCES

[1] Rothan HA, Byrareddy SN. The epidemiology and pathogenesis of coronavirus disease (COVID-19) outbreak. *Journal of Autoimmunity* 2020;102433.

[2] Bogoch II, Watts A, Thomas-Bachli A, Huber C, Kraemer MUG, Khan K. Pneumonia of unknown aetiology in Wuhan, China: potential for international spread via commercial air travel. *J Travel Med* 2020 Mar 13;27(2).

[3] Surveillances V. The Epidemiological Characteristics of an Outbreak of 2019 Novel Coronavirus Diseases (COVID-19)ГÇöChina, 2020. *China CDC Weekly* 2020;2(8):113-22.

[4] WHO Emergency Committee. Statement on the second meeting of the International Health Regulations (2005) *Emergency Committee regarding the outbreak of novel coronavirus (COVID-19)*. 2020.

[5] Wu Z, McGoogan JM. Characteristics of and important lessons from the coronavirus disease 2019 (COVID-19) outbreak in China: summary of a report of 72 314 cases from the Chinese Center for Disease Control and Prevention. *JAMA* 2020.

[6] World Health Organization. *WHO Director-General's opening remarks at the media briefing on COVID-19*-11 March 2020. Geneva, Switzerland 2020.

[7] Lu R, Zhao X, Li J, Niu P, Yang B, Wu H, et al. Genomic characterisation and epidemiology of 2019 novel coronavirus: implications for virus origins and receptor binding. *Lancet* 2020 Feb 22;395(10224):565-74.

[8] Paraskevis D, Kostaki EG, Magiorkinis G, Panayiotakopoulos G, Sourvinos G, Tsiodras S. Full-genome evolutionary analysis of the novel corona virus (2019-nCoV) rejects the hypothesis of emergence as a result of a recent recombination event. *Infection, Genetics and Evolution* 2020;79:104212.

[9] Chen N, Zhou M, Dong X, Qu J, Gong F, Han Y, et al. Epidemiological and clinical characteristics of 99 cases of 2019 novel coronavirus pneumonia in Wuhan, China: a descriptive study. *Lancet* 2020 Feb 15;395(10223):507-13.

[10] Zhang H, Kang Z, Gong H, Xu D, Wang J, Li Z, et al. The digestive system is a potential route of 2019-nCov infection: a bioinformatics analysis based on single-cell transcriptomes. *BioRxiv* 2020.

[11] van Doremalen N, Bushmaker T, Morris DH, Holbrook MG, Gamble A, Williamson BN, et al. Aerosol and Surface Stability of SARS-CoV-2 as Compared with SARS-CoV-1. *New England Journal of Medicine* 2020.

[12] Huang C, Wang Y, Li X, Ren L, Zhao J, Hu Y, et al. Clinical features of patients infected with 2019 novel coronavirus in Wuhan, China. *Lancet* 2020 Feb 15;395(10223):497-506.

[13] Wang Z, Yang B, Li Q, Wen L, Zhang R. Clinical Features of 69 Cases with Coronavirus Disease 2019 in Wuhan, China. *Clin Infect Dis* 2020 Mar 16.

[14] Yasri S, Wiwanitkit V. Clinical features in pediatric COVID-19. *Pediatr Pulmonol* 2020 Mar 20.

[15] Wan S, Xiang Y, Fang W, Zheng Y, Li B, Hu Y, et al. Clinical Features and Treatment of COVID-19 Patients in Northeast Chongqing. *J Med Virol* 2020 Mar 21.

[16] Liu Y, Li J, Feng Y. Critical care response to a hospital outbreak of the 2019-nCoV infection in Shenzhen, China. *Crit Care* 2020 Feb 19;24(1):56.

[17] Cheng Z, Lu Y, Cao Q, Qin L, Pan Z, Yan F, et al. Clinical Features and Chest CT Manifestations of Coronavirus Disease 2019 (COVID-19) in a Single-Center Study in Shanghai, China. *AJR Am J Roentgenol* 2020 Mar 14;1-6.

[18] Xie J, Tong Z, Guan X, Du B, Qiu H, Slutsky AS. Critical care crisis and some recommendations during the COVID-19 epidemic in China. *Intensive Care Med* 2020 Mar 2.

[19] Ruan Q, Yang K, Wang W, Jiang L, Song J. Clinical predictors of mortality due to COVID-19 based on an analysis of data of 150 patients from Wuhan, China. *Intensive care medicine* 2020;1-3.

[20] World Health Organization (Laboratory testing for 2019 novel coronavirus (2019-nCoV) in suspected human cases. *Interim guidance*. Geneva: WHO; 17 Jan 2020.

[21] Shi H, Han X, Jiang N, Cao Y, Alwalid O, Gu J, et al. Radiological findings from 81 patients with COVID-19 pneumonia in Wuhan, China: a descriptive study. *The Lancet Infectious Diseases* 2020.

[22] Lu H. Drug treatment options for the 2019-new coronavirus (2019-nCoV). *Biosci Trends*. 2020. 2020.

[23] Chen N, Zhou M, Dong X, Qu J, Gong F, Han Y, et al. Epidemiological and clinical characteristics of 99 cases of 2019 novel coronavirus pneumonia in Wuhan, China: a descriptive study. *The Lancet* 2020;395(10223):507-13.

[24] Wang M, Cao R, Zhang L, Yang X, Liu J, Xu M, et al. Remdesivir and chloroquine effectively inhibit the recently emerged novel coronavirus (2019-nCoV) *in vitro*. *Cell research* 2020;30(3):269-71.

[25] Toots M, Yoon JJ, Cox RM, Hart M, Sticher ZM, Makhsous N, et al. Characterization of orally efficacious influenza drug with high resistance barrier in ferrets and human airway epithelia. *Science translational medicine* 2019;11(515).

[26] Gautret P, Lagier JC, Parola P, Meddeb L, Mailhe M, Doudier B, et al. Hydroxychloroquine and azithromycin as a treatment of COVID-19: results of an open-label non-randomized clinical trial. *International Journal of Antimicrobial Agents* 2020;105949.

Chapter 2

EPIDEMIOLOGY OF COVID-19

INTRODUCTION

Since the emergence of the 2019 novel coronavirus (2019- nCoV) infection in Wuhan city of China, in December 2019 [1], it has spread rapidly across China and then to the other countries [2]. At the time of writing this chapter (29-3-2020) more than 600000 people in the world are infected with this virus and more than 30000 have succumbed to the disease.

Initially link was reported between the outbreak and the local fish and wild animal market in Wuhan, where several types of wild animals were sold. This market was considered as the starting point of the infection. This inference was made on the basis that most of the initial cases were either working or visiting this market. A possible animal-to-human transmission was thought of. Studies later have demonstrated genetic similarity between the human virus and those obtained from bats sold in the market [3]. The market was closed down for the same reason.

For investigation purpose full-length genome sequences of five patients of novel corona virus was done, at an early stage of the outbreak in China. The sequences were compared to those of bats and showed that, these were identical and share 79.6% sequence identity to SARS-CoV. Furthermore, they showed that that 2019-nCoV is 96% identical at the whole-genome level to a bat coronavirus. Pair wise protein sequence analysis was done with seven conserved non-structural proteins domains and showed that this virus belongs to the species of SARSr-CoV [3]. They also confirmed that 2019-nCoV uses the same cell entry receptor-angiotensin converting enzyme II (ACE2)- as SARS-CoV [3].

Later on studies have demonstrated human-to-human transmission as the major mode of spread of SARS-CoV-2.It spreads through droplets or direct contact [4]. Moreover, presumed hospital-related transmission of SARS-CoV-2 was proved to be as high as in 41% of patients in one study (5). The epidemiological data in Wuhan can be seen in Table-1 [5, 6, 7].

Based on the evidence of a rapidly increasing incidence of infection in the community, the role of transmission by asymptomatic carriers was entertained [8]. High alert for a epidemic spread in China was Given by the health authorities in china [8].

Measures were taken to ensure early detection, early isolation, early treatment, adequate medical supplies, admission to designated corona hospitals, and comprehensive therapeutic strategy to help in control of spread. Collaborative efforts were implemented to combat the novel corona virus, focusing on both persistent strict domestic interventions and vigilance against exogenous imported cases [9].

In addition, it was reported that high transmission rate of SARS-CoV-2, frequent global travel could further enhance its worldwide spread [10]. On 30 January 2020, the WHO declared the COVID-19 outbreak as the sixth public health emergency of international concern [11].

Table 1. Demographic data, underlying medical conditions, clinical manifestations and laboratory findings from three studies of 278 patients with SARS-CoV-2 pneumonia in Wuhan

	Huang et al. (n = 41)	Chen et al. (n = 99)	Wang et al. (n = 138)
Study site	Wuhan local health authority	Wuhan Jinyintan Hospital	Zhongnan Hospital of Wuhan University
Age (years)	49 (41–58)	55.5 (13.1)	56 (42–68)
≥65 years	6 (14.6)	NA	NA
Sex			
Male	30 (73.2)	67 (67.7)	75 (54.3)
Female	11 (26.8)	32 (32.3)	63 (45.7)
Presumed hospital-related infection	NA	NA	57 (41.3)
Healthcare worker	NA	NA	40 (29.0)
Any co-morbidity	13 (31.7)	50 (51.5)	64 (46.4)
Co-morbidities			
Cardiovascular disease	6 (14.6)	40 (40.4)	20 (14.5)
Hypertension	6 (14.6)	NA	43 (31.2)
Diabetes	8 (19.5)	12 (12.1)	14 (10.1)
Respiratory disease	1 (2.4)	1 (1.0)	4 (2.9)
Malignancy	1 (2.4)	1 (1.0)	10 (7.2)
Chronic kidney disease	NA	NA	4 (2.9)
Chronic liver disease	1 (2.4)	NA	4 (2.9)

The earlier infections like H1N1 (2009), polio (2014), Ebola in West Africa (2014), Zika (2016) and Ebola in the Democratic Republic of Congo (2019) were also declared as public health emergency of international concern but none of these had spread rate as alarming as COVID-19. Therefore, it was decided that health workers, governments and the public need to co-operate globally to prevent the spread of this infection. On 11th of March COVID-19 was declared as a pandemic by WHO [12, 13].

THE CHINA EPIDEMIC

In late January the Chinese Center for Disease Control and Prevention (CCDC) issued an epidemic update and risk assessment of COVID-19 [8]. The CCDC document described in detail, about the causative agent epidemiology, clinical manifestations, diagnosis, treatment, and public prevention measures. It also provided practical guidance for people to protect themselves from the infection, (issued on January 27, 2020). It also carried a warning that travelers should avoid all nonessential travel to China [8].

A study was conducted to explore the first 72,314 cases of COVID-19 in China, in 40 days between December 31, 2019 (first recognition of the outbreak) to the end of the study period on February 11, 2020 and concluded that corona virus is highly contagious [14]. It has spread Very rapidly from Wuhan city, to the entire country within a matter of 30 days. Moreover, it has achieved such far-reaching effects in spite of extreme response measures like complete shutdown and isolation of entire cities, cancellation of Chinese new year celebrations, shut down of schools and work, massive mobilization of health care workers, as well as military medical units, and rapid construction of corona care hospitals [14]. Mild symptoms were noticed in 81% of COVID-19 cases. Overall fatality was assessed to be 2.3%. Among the 1,023 deaths, maximum were reported in above 60 age group, and the death happened mainly among people having pre-existing, conditions like hypertension, cardiovascular disease, and diabetes. Table 2 [14].

EPIDEMIC TO PANDEMIC

The spread of COVID-19 is becoming uncontrollable and has already exceeded the necessary epidemiological criteria for it to be declared a pandemic [15, 16], having infected more than 600 000 people in more than190 countries. The death rate has exceeded 33000as on 29th march 2020.

Figure-2 shows the statistics of infection as on 29th March 2020.

The maximum affected countries include US, Italy, Spain China France and Iran. Therefore, a coordinated and efficient global response is desperately needed to prepare health systems to meet this unprecedented challenge.

Country, Other	Total Cases	New Cases	Total Deaths	New Deaths	Total Recovered	Active Cases	Serious, Critical	Tot Cases/ 1M pop	Deaths/ 1M pop	1st case
World	691,493	+28,411	33,034	+2,177	147,354	511,105	25,476	88.7	4.2	Jan 10
USA	125,308	+1,730	2,246	+25	3,532	119,530	2,666	379	7	Jan 20
Italy	97,689	+5,217	10,779	+756	13,030	73,880	3,906	1,616	178	Jan 29
China	81,439	+45	3,300	+5	75,448	2,691	742	57	2	Jan 10
Spain	78,799	+5,564	6,606	+624	14,709	57,484	4,165	1,685	141	Jan 30
Germany	58,247	+552	455	+22	8,481	49,311	1,581	695	5	Jan 26
Iran	38,309	+2,901	2,640	+123	12,391	23,278	3,206	456	31	Feb 18
France	37,575		2,314		5,700	29,561	4,273	576	35	Jan 23
UK	19,522	+2,433	1,228	+209	135	18,159	163	288	18	Jan 30
Switzerland	14,829	+753	300	+36	1,595	12,934	301	1,713	35	Feb 24
Netherlands	10,866	+1,104	771	+132	3	10,092	914	634	45	Feb 26
Belgium	10,836	+1,702	431	+78	1,359	9,046	867	935	37	Feb 03
S. Korea	9,583	+105	152	+8	5,033	4,398	59	187	3	Jan 19

Figure 1. COVID 19 statistics.

Although the containment measures implemented in China have reduced new cases by more than 90%, this reduction has not been achieved in many developed countries, including US, Italy, spain, UK and Iran.

MEASURES TO CONTROL THE PANDEMIC – LESSONS LEARNED

The key factors that influence course of an epidemic are poorly understood at present for COVID-19. The basic reproduction number (R0), is the mean number of secondary cases infected by one primary case. It determines the total number of people will get infected, and determines the area under the epidemic curve. For an epidemic to get controlled, the value of R0 must be less than unity in value. A simple calculation gives the fraction likely to get infected without mitigation. This fraction is roughly 1–1/R0. With R0 values for COVID-19 in China around 2·5 in the early stages of the epidemic [17].

Voluntary social distancing by individuals and communities will have an Impacts, and mitigation efforts, such as the measures taken by China, greatly reduce transmission. As an epidemic progresses, the effective reproduction number (R) declines until it falls below one. Any epidemic will reach a peak then declines, either due to the limitation of people susceptible to infection or secondary to control measures imposed. The two factors that determines the likely duration of the epidemic are the incubation period and mean duration of infectiousness in a patient.

The incubation period for COVID-19 is about 5–6 days and may go up to 14 days [18]. There might be considerable pre symptomatic infectious cases and a large number of asymptomatic cases of COVID-19. Estimates suggest that about 80% of people with COVID-19 have mild or asymptomatic disease, 14% have severe disease, and 6% are critically ill, so the symptom-based control is unlikely to be efficient as spread occurs mainly through asymptomatic cases. The epidemiologists has a prominent role is helping policy makers to minimize the morbidity and mortality, avoiding an epidemic peak that is beyond the capacity health-care services, keeping the impact on the economy to manageable levels, and flattening the epidemic curve (Figure 3), till an effective vaccine or antiviral drug therapies become available.

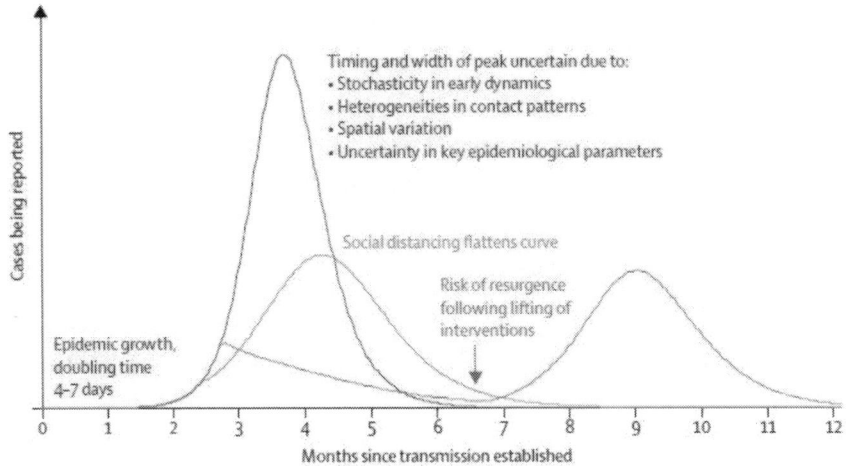

Figure 2. Simulations of a transmission model of COVID-19.

REFERENCES

[1] Lu H, Stratton CW, Tang Y. Outbreak of Pneumonia of Unknown Etiology in Wuhan China: the Mystery and the Miracle. *Journal of Medical Virology*.

[2] Lai CC, Wang CY, Wang YH, Hsueh SC, Ko WC, Hsueh PR. Global epidemiology of coronavirus disease 2019: disease incidence, daily cumulative index, mortality, and their association with country healthcare resources and economic status. *Int J Antimicrob Agents* 2020 Mar 18;105946.

[3] Zhou P, Yang XL, Wang XG, Hu B, Zhang L, Zhang W, et al. A pneumonia outbreak associated with a new coronavirus of probable bat origin. *Nature* 2020;1-4.

[4] Li Q, Guan X, Wu P, Wang X, Zhou L, Tong Y, et al. Early transmission dynamics in Wuhan, China, of novel coronavirusΓÇôinfected pneumonia. *New England Journal of Medicine* 2020.

[5] Wang D, Hu B, Hu C, Zhu F, Liu X, Zhang J, et al. Clinical characteristics of 138 hospitalized patients with 2019 novel coronavirusΓÇôinfected pneumonia in Wuhan, China. *JAMA* 2020.

[6] Chen N, Zhou M, Dong X, Qu J, Gong F, Han Y, et al. Epidemiological and clinical characteristics of 99 cases of 2019 novel coronavirus pneumonia in Wuhan, China: a descriptive study. *The Lancet* 2020;395(10223):507-13.

[7] Huang C, Wang Y, Li X, Ren L, Zhao J, Hu Y, et al. Clinical features of patients infected with 2019 novel coronavirus in Wuhan, China. *Lancet* 2020 Feb 15;395(10223):497-506.

[8] Biscayart C, Angeleri P, Lloveras S, Chaves TDSS, Schlagenhauf P, Rodr+¡guez-Morales AJ. The next big threat to global health? 2019 novel coronavirus (2019-nCoV): What advice can we give to travellers?-Interim recommendations January 2020, from the Latin-American society for Travel Medicine (SLAMVI). *Travel medicine and infectious disease* 2020;33:101567.

[9] Fang Y, Nie Y, Penny M. Transmission dynamics of the COVID-19 outbreak and effectiveness of government interventions: A data-driven analysis. *J Med Virol* 2020 Mar 6.

[10] Liu SL, Saif L. Emerging Viruses without Borders: The Wuhan Coronavirus. *Viruses* 2020 Jan 22;12(2).

[11] Sohrabi C, Alsafi Z, O'Neill N, Khan M, Kerwan A, Al-Jabir A, et al. World Health Organization declares global emergency: A

review of the 2019 novel coronavirus (COVID-19). *Int J Surg* 2020 Feb 26;76:71-6.

[12] Cucinotta D, Vanelli M. WHO Declares COVID-19 a Pandemic. *Acta Biomed* 2020 Mar 19;91(1):157-60.

[13] World Health Organization. *WHO Director-General's opening remarks at the media briefing on COVID-19*-11 March 2020. Geneva, Switzerland 2020.

[14] Surveillances V. The Epidemiological Characteristics of an Outbreak of 2019 Novel Coronavirus Diseases (COVID-19)ГÇöChina, 2020. *China CDC Weekly* 2020;2(8):113-22.

[15] World Health Organization. Coronavirus disease 2019 (ГÇÄ COVID-19) *ГÇÄ: situation report*, 51. 2020.

[16] Callaway E. Time to use the p-word? Coronavirus enter dangerous new phase. *Nature* 2020;579:12.

[17] Imai N, Cori A, Dorigatti I, Baguelin M, Donnelly CA, Riley S, et al. *Report 3: transmissibility of 2019-nCov*. Imperial College London. 2020.

[18] Thompson R. Pandemic potential of 2019-nCoV. *The Lancet Infectious Diseases* 2020;20(3):280.

Chapter 3

CLINICAL PRESENTATION OF COVID-19

INTRODUCTION

The symptoms of COVID-19 starts to appear after an incubation period, which on an average is about 5 days of exposure to an infected individual [1]. Sometimes the incubation period may go upto14 days and this is the rationale behind the fourteen days quarantine period advised by the health authorities [1]. Patients with good immunity and younger patients have a longer incubation period but in the elderly the incubation period is shorter. The most common symptoms by which the infection starts is with fever, headache and cough. Most patients have a mild course and are often unnoticed by the individual. Mild cases constitute about 70-80% [2]. About 5% of cases become critical and mostly occur in elderly or individuals with other co-morbid conditions [3].

CLINICAL FEATURES

Patients infected with the virus start to develop upper respiratory symptoms resembling flu [4-6]. The most frequent symptoms of illness at the onset are fever (82.1%), cough (45.8%), fatigue (26.3%), dyspnea (6.9%) and headache (6.5%). Severe cases had high incidence of dyspnea (32.6%) [2, 7].

The median age group in which COVID-19 presents is 51 years and has almost equal male to female incidence [8]. The median duration from onset of symptoms to hospitalization was 4 (range 2-7) days in symptomatic individuals [8]. About (86.3%) patients were discharged after 16 days (range 12-20 days) of hospitalization [8]. The estimated median duration of fever in all the patients with fever was 10 days (95 confidence intervals being 8-11 days [8]. Patients who require intensive care units (ICU) admissions had significantly prolonged fever as compared to those not in ICU (31 days Vs 9 days after onset of symptoms respectively, $P < 0.0001$) [8]. Less common symptoms were sputum production [28%], haemoptysis [5%] myalgia, sore throat and diarrhea [9].

COVID-19 pneumonia presents with chest pain, difficulty in breathing, confusion, difficulty in walking, bluish discoloration of lips and sometimes syncope. On evaluation these patients will have abnormal chest x-ray and Chest CT findings.

With regard to the cardiovascular system, ECG often show variations of arrhythmia or features of myocarditis [10]. Angiotensin - converting enzyme 2 (ACE2) is a membrane bound amino peptidase that plays a major role in the cardiovascular and immune systems [11]. ACE2 is involved in heart function and the development of diabetes mellitus and hypertension [12]. ACE2 acts as a functional receptor for corona viruses including SARS-CoV and SARS- CoV-2. COVID-19 infection is triggered by binding of the spike protein of the virus to ACE2, which is highly expressed in the lungs and the heart. SARS-CoV-2 mainly invades alveolar epithelial cells, resulting in respiratory symptoms and ARDS [12]. Up to two- third of patients have anosmia and a significant numbers have ageusia. In the nervous system ischemic stroke, hemorrage and facial palsy are common presentation. Vascular complications are thought to be due to endothelitis and subsequent thrombosis occuring in various organs like brain ,heart, kidney, liver and peripheral vasculature. Rare cases of AIDP also has been reported in covid.

ARDS AND MULTI-ORGAN DYSFUNCTION

This develops usually in 7-10 days of infection in approximately 5% of cases. Apparently normal patient suddenly deteriorate with dyspnoea and sudden hypoxia. These patients may improve with oxygen but many will require mechanical ventilation. Renal involvement predicts poor prognosis in such patients. The cytokine storm syndrome is thought to be the main reason for sudden worsening of COVID-19 [13]. Respiratory failure, septic shock and/or multiple organ dysfunction (MOD) indicate grave outcome.

CLINICAL CLASSIFICATION

Clinical classification [14] is often used to help in management of COVID-19 cases.

Mild Disease

Symptoms are mild and no evidence of pneumonia.

Moderate Disease

Patient has fever, respiratory symptoms and evidence of pneumonia on chest imaging.

Severe Disease

Patients with a respiratory rate more than 30, Spo2 less than 93% and evidence of Pneumonia.

Critical Disease

The diagnosis requires clinical and ventilatory criteria. This syndrome is suggestive of a serious new-onset respiratory failure or for worsening of an already identified respiratory picture. Different forms of ARDS are distinguished based on the degree of hypoxia.

ASYMPTOMATIC CASES

Asymptomatic infection, screened from close contacts has demonstrated the transmission potential of infection through asymptomatic carriers [15]. Many of these patients have later on developed fever and other symptoms on follow up. Some have developed abnormal chest CT. A few will develop full-fledged clinical features of COVID-19 on follow up. History of anosmia and ageusia is often helpful in otherwise asymptomatic individual.

MORTALITY AND PREDICTORS

Predictors of a fatal outcome in COVID-19 cases include advanced age, the presence of underlying diseases, the presence of secondary infection and elevated inflammatory indicators in the blood. Comorbidities that is associated with bad prognosis include diabetes mellitus, ischemic heart disease, hypertension, COPD and immunosuppression [16].

REFERENCES

[1] Li Q, Guan X, Wu P, Wang X, Zhou L, Tong Y, et al. Early transmission dynamics in Wuhan, China, of novel coronavirusGÇôinfected pneumonia. *New England Journal of Medicine* 2020.

[2] Tian S, Hu N, Lou J, Chen K, Kang X, Xiang Z, et al. Characteristics of COVID-19 infection in Beijing. *J Infect* 2020 Apr;80(4):401-6.

[3] Zhou F, Yu T, Du R, Fan G, Liu Y, Liu Z, et al. Clinical course and risk factors for mortality of adult inpatients with COVID-19 in Wuhan, China: a retrospective cohort study. *Lancet* 2020 Mar 11.

[4] Wang Z, Chen X, Lu Y, Chen F, Zhang W. Clinical characteristics and therapeutic procedure for four cases with 2019 novel coronavirus pneumonia receiving combined Chinese and Western medicine treatment. *Biosci Trends* 2020 Mar 16;14(1):64-8.

[5] Wang D, Hu B, Hu C, Zhu F, Liu X, Zhang J, et al. Clinical characteristics of 138 hospitalized patients with 2019 novel coronavirusGÇôinfected pneumonia in Wuhan, China. *JAMA* 2020.

[6] Zhang JJ, Dong X, Cao YY, Yuan YD, Yang YB, Yan YQ, et al. Clinical characteristics of 140 patients infected with SARS-CoV-2 in Wuhan, China. *Allergy* 2020 Feb 19.

[7] Wu Z, McGoogan JM. Characteristics of and important lessons from the coronavirus disease 2019 (COVID-19) outbreak in China: summary of a report of 72 314 cases from the Chinese Center for Disease Control and Prevention. *JAMA* 2020.

[8] Chen J, Qi T, Liu L, Ling Y, Qian Z, Li T, et al. Clinical progression of patients with COVID-19 in Shanghai, China. *J Infect* 2020 Mar 19.

[9] Huang C, Wang Y, Li X, Ren L, Zhao J, Hu Y, et al. Clinical features of patients infected with 2019 novel coronavirus in Wuhan, China. *Lancet* 2020 Feb 15;395(10223):497-506.

[10] Zeng JH, Liu YX, Yuan J, Wang FX, Wu WB, Li JX, et al. First Case of COVID-19 Infection with Fulminant Myocarditis Complication: *Case Report and Insights*. 2020.

[11] Zheng YY, Ma YT, Zhang JY, Xie X. COVID-19 and the cardiovascular system. *Nature Reviews Cardiology* 2020;1-2.

[12] Turner AJ, Hiscox JA, Hooper NM. ACE2: from vasopeptidase to SARS virus receptor. *Trends in pharmacological sciences* 2004;25(6):291-4.

[13] Mehta P, McAuley DF, Brown M, Sanchez E, Tattersall RS, Manson JJ. COVID-19: consider cytokine storm syndromes and immunosuppression. *The Lancet* 2020.

[14] Wu Z, McGoogan JM. Characteristics of and important lessons from the coronavirus disease 2019 (COVID-19) outbreak in China: summary of a report of 72 314 cases from the Chinese Center for Disease Control and Prevention. *JAMA* 2020.

[15] Hu Z, Song C, Xu C, Jin G, Chen Y, Xu X, et al. Clinical characteristics of 24 asymptomatic infections with COVID-19 screened among close contacts in Nanjing, China. *Science China Life Sciences* 2020;1-6.

[16] Ruan Q, Yang K, Wang W, Jiang L, Song J. Clinical predictors of mortality due to COVID-19 based on an analysis of data of 150 patients from Wuhan, China. *Intensive care medicine* 2020;1-3.

Chapter 4

THE VIRUS

INTRODUCTION

Severe acute respiratory syndrome corona virus 2 (SARS-CoV-2) is an enveloped, single positive stranded RNA virus, which belongs to the subfamily Coronavirinae [1]. A close genetic similarity has been identified between this virus and corona virus isolated from bats [2]. Hence a zoonotic origin, followed by human to human spread is the most likely mode of transmission of COVID-19 [3]. As the site of origin of the virus is Wuhan, it is sometimes referred to as the "Wuhan virus" or "Wuhan corona virus". But World Health Organization does not encourage use of names based upon locations, hence a better term will be to use - virus responsible for COVID-19 [4]. An intermediate animal reservoir such as a pangolin may act as an intermediate host in infecting humans [4].

MORPHOLOGY

The genome, of the virus causing COVID-19 has a length of about 26 to 32 kilobases and is probably the largest viral RNA known [5].

Corona viruses (CoVs) are the largest group of viruses belonging to the *Nidovirales* order which includes *Coronaviridae, Arteriviridae, Mesoniviridae and Roniviridae* families. The *Coronavirinae* comprise one of two subfamilies in the *Coronaviridae* family, with the other being the *Torovirinae*. The *Coronavirinae* are further subdivided into four genera, the alpha, beta, gamma and delta corona viruses. The viruses were initially sorted into these genera based on serology but are now divided by phylogenetic clustering.

All viruses in the *Nidovirales* order are enveloped, non-segmented positive-sense RNA viruses. They all contain very large genomes for RNA viruses, with some viruses having the largest identified RNA genomes, containing up to 33.5 kilobase (kb) genomes. Other common features within the *Nidovirales* order include: (1) a highly conserved genomic organization, with a large replicase gene preceding structural and accessory genes; (2) expression of many non-structural genes by ribosomal frame shifting; (3) several unique or unusual enzymatic activities encoded within the large replicase–transcriptase polyprotein; and (4) expression of downstream genes by synthesis of 3′ nested sub-genomic mRNAs. In fact, the *Nidovirales* order name is derived from these nested 3′ mRNAs as *nido* is Latin for "nest." The major differences within the Nidovirus families are in the number, type and sizes of the structural proteins. These differences cause significant alterations in the structure and morphology of the nucleocapsids and virions [5].

The 2019-nCoV is included in lineage B β-COVs, subgenus Sarbecovirus, which possesses a single-stranded positive-sense RNA surrounded by an envelope. The RNA genome includes 29,891 nucleotides (GenBank no. MN908947), encoding 9860 amino acids and this is arranged in the order of 5 _ UTR - replicase (orf1a/b) - Spike(*S*) – Envelope (E) - Membrane (M) -Nucleocapsid (N) −3 UTR, in which S, E, M, N encodes the structural proteins [6] (Figure 1).

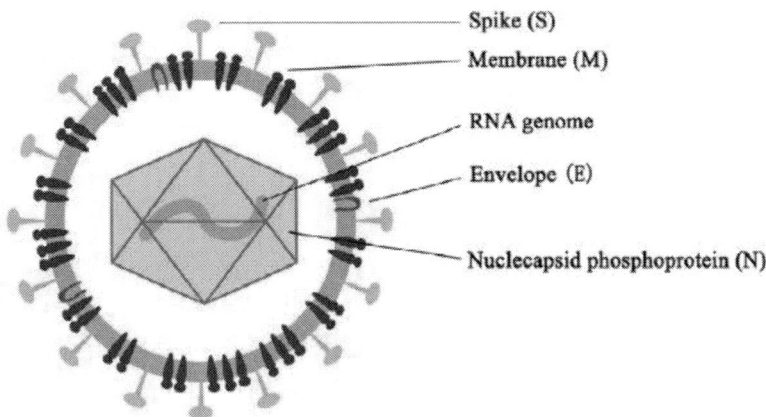

Figure 1.

The genome of the 2019-nCoV consists of six major functional open reading frames (ORFs), including ORF1a/b, S, E, M, N and several other accessory genes, such as ORF3b and OFR8. Replicase polyproteins pp1a and pp1ab, which are ORF1a/b would be proteolytic cleaved into 16 non-structural proteins (nsps) which are involved in the transcription and replication of the virus [6].

In addition, the ORF3b encoded a completely novel short protein without an exact function. The new ORF8 likely encodes a secreted protein formed by an alpha-helix, followed by beta sheets that contain six strands that have no known functional domain or motif similarity [6]. The S gene of the 2019-nCoV has less than 75% sequence identity to those of the two CoVs, bat SARS-like CoVs (SL-CoVZXC21 and ZC45) and human SARS-CoV [2, 6]. Similarly, the spike glycoprotein encoded by the S genes of the 2019-nCoV was longer than that of the SARS-CoV. The spike protein, composed of S1 and S2 domain, was crucial in determining host tropism and transmission capacity through the mediation of receptor binding and membrane fusion. Among these, the S2 subunit of the 2019-nCoV is highly conserved and has a 99% identity with that of SARS-CoV [6]. The receptor-binding domain is commonly located in the C-terminal domain of S1 to directly contact the human receptor. Although, the S1 domain of the 2019-nCoV only has approximately 70% identity with SARS-CoV, homology modeling revealed that the 2019-nCoV has a similar receptor-binding domain structure to that of SARS- CoV [7], Notably, Zhou et al. found that the 2019-nCoV just like SARS-Cov may also use ACE2 as an entry receptor in the ACE2- expressing cells, and the majority of which are type II alveolar cells (AT2) in human lungs [2, 8]. Therefore, further investigations are needed to determine whether ACE2 targeting drugs would be effective for the treatment of 2019-nCoV [8]. Since the genomic sequences of the 2019-nCoV obtained from different patients were extremely similar to each other, which exhibited more than 99. 9% sequence identity, it could be reasonably considered that the 2019-nCoV originated from one source rather than a mosaic and could be detected relatively rapidly. However, mutations need to be constantly monitored when the virus is transmitting to an increasing number of individuals.

VIRAL ENTRY AND MULTIPLICATION

Corona virus uses its spike glycoprotein (S), to bind its receptor, (Figure 2) and mediate membrane fusion and virus entry. Each monomer of trimeric S protein is about 180 kDa, and contains two subunits, S1 and S2, mediating attachment and membrane fusion, respectively. In the structure, N- and C- terminal portions of S1 fold as two independent domains, N-terminal domain (NTD) and C-terminal domain (C-domain). The receptors for SARS-CoV and MERS-CoV are human angiotensin-converting enzyme 2 (hACE2) [9].

The Virus

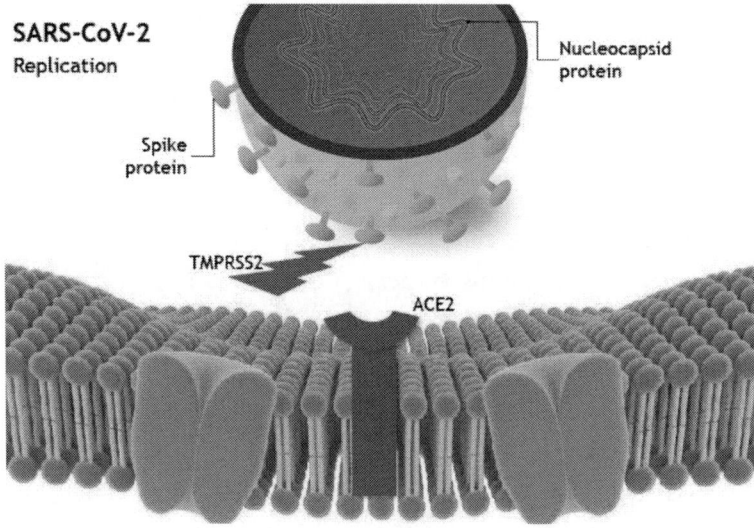

Figure 2.

Studies have reported or predicted human ACE2 usage of 2019-nCoV in a similar way to SARS-CoV mainly based on the coronavirus spike (S) glycoproteins [10]. The binding of a receptor expressed by host cells is the first step of viral infection followed by fusion with the cell membrane. It is reasoned that the lung epithelial cells are the primary target of the virus. Thus, it has been reported that human-to-human transmissions happens via the cellular receptor which has been identified as angiotensin converting enzyme 2 (ACE2) receptor. ACE 2 receptors are also seen on endothelial surfaces of blood vessels. Hence this receptor could be a possible target that can be used in future drug development against COVID-19.

REFERENCES

[1] Fehr AR, Perlman S. Coronaviruses: an overview of their replication and pathogenesis. *Coronaviruses.* Springer; 2015. p. 1-23.

[2] Zhou P, Yang XL, Wang XG, Hu B, Zhang L, Zhang W, et al. A pneumonia outbreak associated with a new coronavirus of probable bat origin. *Nature* 2020;579(7798):270-3.

[3] Perlman S. Another decade, another coronavirus. 2020. *Mass Medical Soc.*

[4] World Health Organization. World Health Organization best practices for the naming of new human infectious diseases. *World Health Organization*; 2015.

[5] Fehr AR, Perlman S. Coronaviruses: an overview of their replication and pathogenesis. *Coronaviruses.* Springer; 2015. p. 1-23.

[6] Chan JF-W, Kok KH, Zhu Z, Chu H, To KK-W, Yuan S, et al. Genomic characterization of the 2019 novel human-pathogenic coronavirus isolated from a patient with atypical pneumonia after visiting Wuhan. *Emerging Microbes & Infections* 2020;9(1):221-36.

[7] Lu R, Zhao X, Li J, Niu P, Yang B, Wu H, et al. Genomic characterisation and epidemiology of 2019 novel coronavirus: implications for virus origins and receptor binding. *The Lancet* 2020;395(10224):565-74.

[8] Zhao Y, Zhao Z, Wang Y, Zhou Y, Ma Y, Zuo W. Single-cell RNA expression profiling of ACE2, the putative receptor of Wuhan 2019-nCov. *BioRxiv* 2020.

[9] Ou X, Liu Y, Lei X, Li P, Mi D, Ren L, et al. Characterization of spike glycoprotein of SARS-CoV-2 on virus entry and its immune cross-reactivity with SARS-CoV. *Nature Communications* 2020;11(1):1-12.

[10] Lu G, Hu Y, Wang Q, Qi J, Gao F, Li Y, et al. Molecular basis of binding between novel human coronavirus MERS-CoV and its receptor CD26. *Nature* 2013;500(7461):227-31.

Chapter 5

PATHOGENESIS AND PATHOLOGY

INTRODUCTION

The reverse genetic system has revolutionized the study of viral replication, cell biology and pathogenesis of corona virus. The renewed interest in this class of viruses happened following the SARS epidemic and research funding in study of these viruses. The impressive speed with which the SARS-associated virus was identified and the genome sequenced was made possible by the data accumulated previously on the other members of the same family. The novel corona virus was studied in the early stages of the pandemic and genetic sequencing was made available on the basis of experience gained on SARS infection. This has made the study of pathogenesis of COVID-19 virus easier in the current pandemic of novel corona virus.

PATHOGENESIS OF COVID-19

The virus enters the host cell with the help of Corona virus S protein [1]. The spike glycoprotein on the envelope binds to its cellular receptor, ACE2. The binding of the virus with host cell receptors is a significant determinant for the pathogenesis of infection. Direct membrane fusion between the virus and plasma membrane is the mechanism by which virus enters the cell [1, 2]. After getting into the cell the viral genome begins to replicate [3]. The newly formed envelope glycoprotein is inserted into the membrane of the endoplasmic reticulum or Golgi bodies. The nucleocapsid is formed by the combination of genomic RNA and nucleocapsid protein. Then, viral particles germinate into the endoplasmic reticulum-Golgi intermediate compartment. The Vesicles containing the virus particles in turn fuse with the plasma membrane and the new virus is released [1, 4].

The main pathogenesis of COVID-19 is through attachment to the ACE 2 receptors abundant in the respiratory tract [5]. The virus replicates and subsequently infects other cells in the upper respiratory tract and lung tissue; patients then begin to have clinical symptoms and manifestations. Pathological studies in patients with COVID-19 virus confirmed the presence of the virus in liver tissue, although the viral titer was relatively low because viral inclusions were not observed in liver [6]. Entry into the lung result in pneumonia and ARDS [7]. Severe disease onset might result in death due to massive alveolar damage and progressive respiratory failure, Significantly high blood levels of cytokines and chemokines were observed in patients with COVID-19 infection that included IL1-β, IL1RA, IL7, IL8, IL9, IL10, basic FGF2, GCSF, GMCSF, IFNγ, IP10, MCP1, MIP1α, MIP1β, PDGFB, TNFα, and VEGFA [8]. Some of the severe cases that were admitted to the intensive care unit showed high levels of pro-inflammatory cytokines including IL2, IL7, IL10, GCSF, IP10, MCP1, MIP1α, and TNFα resulting in severe disease [9, 10, 11]. The viral invasion can occur to cardiac tissue resulting features of myocarditis and permanent damage leading to cardiomyopathy. ACE2 receptor are expressed in cholangiocytes, indicating that SARS-CoV-2 might directly bind to ACE2-positive cholangiocytes resulting in alteration of liver function.

GI invasion and viral shedding is known to occur even weeks after COVID-19 [12].

Endothelial Invasion occur in COVID-19 infection and result in high incidence of cardiovascular and cerebrovascular events [13]. ACE2 receptors are widely distributed in endothelial cells. Endothelial dysfunction is a principal determinant of microvascular dysfunction by shifting the vascular equilibrium towards more vasoconstriction with subsequent organ ischaemia, inflammation with associated tissue oedema, and a pro-coagulant state [13]. Thrombosis of arteries in lung, liver and kidneys are common in critically ill COVID-19 cases.

PATHOLOGY OF COVID-19

Biopsy specimen taken from the lung of a patient who died of COVID-19 has shown bilateral diffuse alveolar damage with cellular fibromyxoid exudates. There were evident alveolar damage, including alveolar edema and proteinaceous exudates. There were desquamation of pneumocytes and hyaline membrane formation, indicating acute respiratory distress syndrome [12]. Interstitial mononuclear inflammatory infiltrates, dominated by lymphocytes, were also seen.

Multinucleated (characterized by large nuclei), amphophilic granular cytoplasm, and prominent nucleoli were identified in the intra alveolar spaces, showing viral cytopathic-like changes. No obvious intranuclear or intracytoplasmic viral inclusions were identified. But some have reported presence of viral inclusions in lung specimens [14]. There were syncytial cells with atypical enlarged pneumocytes [15]. There was diffuse thickening of alveolar walls (Figure-1), consisting of proliferating interstitial fibroblasts and type II pneumocyte hyperplasia. Focal fibroblast plug and multinucleated giant cells were seen in the airspaces (Figure2), indicating varying degrees of proliferative phase of diffuse alveolar damage. Some areas had abundant alveolar macrophages along with type II pneumocyte hyperplasia.

The liver, biopsy specimen showed moderate microvesicular steatosis and mild lobular and portal activity. Liver damage in mild cases of COVID-19 is often transient and most often return to normal without any special treatment. However, in cases of severe liver damage, liver protective drugs have been tried [16]. In the cardiac tissue there were a few interstitial mononuclear inflammatory infiltrates, but no significant damage in the heart tissue [14, 17]. Acute hemorrhagic necrotizing encephalopathy has been reported in COVID-19 infections [18]. Increased incidence of hemorrhagic strokes has been communicated by neurologists of Kerala, India during the pandemic (personal communications-unpublished). Hyper coagulopathy with pulmonary thrombosis, increased incidence of ischemic stroke also has been reported.

Presence of viral elements in the endothelial cells and an accumulation of inflammatory cells, with evidence of endothelial and inflammatory cell death has been documented [13]. These findings suggest that SARS-CoV-2 infection facilitates the induction of endotheliitis in several organs as a direct consequence of viral involvement [13]. Pathologically the staining patterns were consistent with apoptosis of endothelial cells and mononuclear cells (Figure 3).

The cytometric analysis of peripheral blood showed reduced peripheral CD4 and CD8 T cells, while their status was hyper activated, as evidenced by the high proportions of HLA-DR (CD4 3·47%) and CD38 (CD8 39·4%) double-positive fractions [14]. The concentration of pro inflammatory CCR6+ Th17 in CD4 T cells were increased. CD8 T cells were found to have high concentrations of cytotoxic granules, in which 31·6% cells were perforin positive, 64·2% cells were granulysin positive, and 30·5% cells were granulysin and perforin double-positive. These findings imply that over activation of T cells, manifested by increase of Th17 and high cytotoxicity of CD8 T cells, resulting in, in part, and the severe immune injury in these patients [14].

Pathogenesis and Pathology

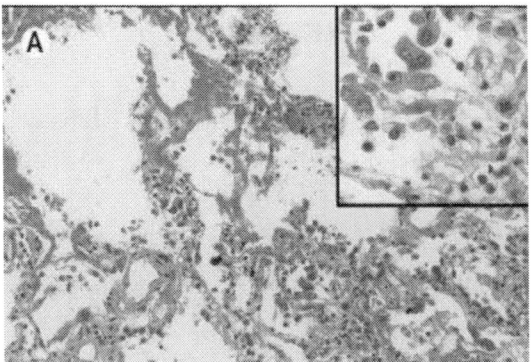

Figure 1.

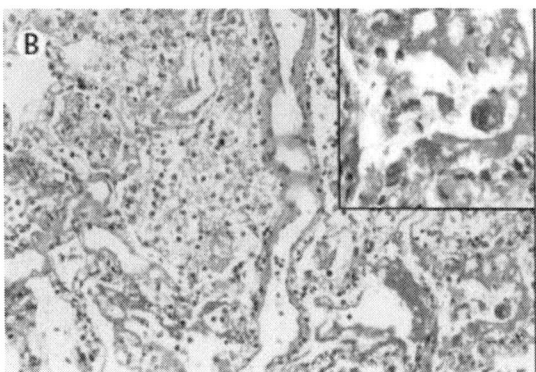

Figure 2.

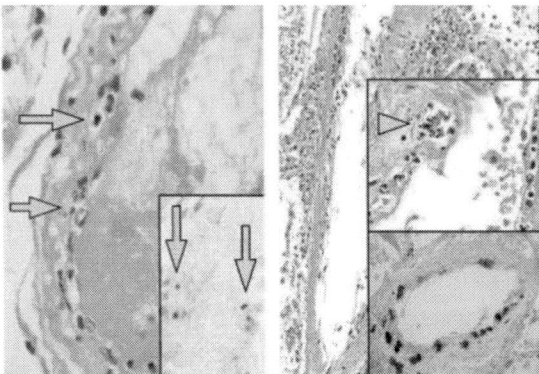

Figure 3.

REFERENCES

[1] de Wit E, van Doremalen N, Falzarano D, Munster VJ. SARS and MERS: recent insights into emerging coronaviruses. *Nature Reviews Microbiology* 2016;14(8):523.

[2] Solomon S, Daniel JS, Sanford TJ, Murphy DM, Plattner GK, Knutti R, et al. Persistence of climate changes due to a range of greenhouse gases. *Proceedings of the National Academy of Sciences* 2010;107(43):18354-9.

[3] Perlman S, Netland J. Coronaviruses post-SARS: update on replication and pathogenesis. *Nature Reviews Microbiology* 2009;7(6):439-50.

[4] Li X, Geng M, Peng Y, Meng L, Lu S. Molecular immune pathogenesis and diagnosis of COVID-19. *Journal of Pharmaceutical Analysis* 2020.

[5] Li W, Moore MJ, Vasilieva N, Sui J, Wong SK, Berne MA, et al. Angiotensin-converting enzyme 2 is a functional receptor for the SARS coronavirus. *Nature* 2003;426(6965):450-4.

[6] Chau T, Lee K, Yao H, Tsang T, Chow T, Yeung Y, et al. SARSGÇÉassociated viral hepatitis caused by a novel coronavirus: report of three cases. *Hepatology* 2004;39(2):302-10.

[7] Seguin Al, Galicier L, Bouthoul D, Lemiale V, Azoulay E. Pulmonary involvement in patients with hemophagocytic lymphohistiocytosis. *Chest* 2016;149(5):1294-301.

[8] Huang C, Wang Y, Li X, Ren L, Zhao J, Hu Y, et al. Clinical features of patients infected with 2019 novel coronavirus in Wuhan, China. *The Lancet* 2020;395(10223):497-506.

[9] Rothan HA, Byrareddy SN. The epidemiology and pathogenesis of coronavirus disease (COVID-19) outbreak. *Journal of Autoimmunity* 2020;102433.

[10] Mehta P, McAuley DF, Brown M, Sanchez E, Tattersall RS, Manson JJ. COVID-19: consider cytokine storm syndromes and immunosuppression. *The Lancet* 2020.

[11] Richardson P, Griffin I, Tucker C, Smith D, Oechsle O, Phelan A, et al. Baricitinib as potential treatment for 2019-nCoV acute respiratory disease. *The Lancet* 2020;395(10223):e30-e31.

[12] Xu Z, Shi L, Wang Y, Zhang J, Huang L, Zhang C, et al. Pathological findings of COVID-19 associated with acute respiratory distress syndrome. *The Lancet respiratory medicine* 2020.

[13] Varga Z, Flammer AJ, Steiger P, Haberecker M, Andermatt R, Zinkernagel AS, et al. Endothelial cell infection and endotheliitis in COVID-19. *The Lancet* 2020.

[14] Xu Z, Shi L, Wang Y, Zhang J, Huang L, Zhang C, et al. Pathological findings of COVID-19 associated with acute respiratory distress syndrome. *The Lancet respiratory medicine* 2020.

[15] Tian S, Hu W, Niu L, Liu H, Xu H, Xiao SY. Pulmonary pathology of early phase 2019 novel coronavirus (COVID-19) pneumonia in two patients with lung cancer. *Journal of Thoracic Oncology* 2020.

[16] Zhang C, Shi L, Wang FS. Liver injury in COVID-19: management and challenges. *The Lancet Gastroenterology & Hepatology* 2020.

[17] Inciardi RM, Lupi L, Zaccone G, Italia L, Raffo M, Tomasoni D, et al. Cardiac Involvement in a Patient With Coronavirus Disease 2019 (COVID-19). *JAMA cardiology* 2020.

[18] Poyiadji N, Shahin G, Noujaim D, Stone M, Patel S, Griffith B. COVID-19GÇôassociated Acute Hemorrhagic Necrotizing Encephalopathy: CT and MRI Features. *Radiology* 2020;201187.

Chapter 6

DIAGNOSIS OF COVID-19

INTRODUCTION

Investigating a case for COVID-19 is based on clinical history, contact history and epidemiological parameters. Regions with evidence of community spread will require wide spread sampling to detect asymptomatic cases as these cases play a key role on community spread [1]. Suspect cases should be screened for the virus with nucleic acid amplification tests (NAAT), such as RT-PCR.A [1], several antibody tests are now available. Rapid antibody test kits can be used in areas of community spread for easy identification of cases and isolation. WHO has putdown definite guidelines to diagnose COVID-19 [2]. The information in this chapter will be predominantly based on these recommendations.

LAB INVESTIGATIONS

Specific Tests for COVID-19

Routine confirmation of cases of COVID-19 is based on detection of unique sequences of virus RNA by NAAT such as real-time reverse transcription Polymerase chain reaction (rRT-PCR). The viral genes targeted usually include the N, E, S and RdRP genes. Specimens from upper respiratory tract like nasopharyngeal and oropharyngeal swabs, nasopharyngeal wash/nasopharyngeal aspirate can be collected. In lower respiratory tract nasopharyngeal lavage or aspirate with the help of bronchoscope is used. Other specimens like sputum, urine [3], stool, tear viral yield is less [4].

In an area with no covid infection a positive NAAT result for at least two different targets on the COVID-19 virus genome, of which at least one target is preferably specific for COVID-19 virus using a validated assay. When there are discordant results, re sampling should be done and, if appropriate, sequencing of the virus from the original specimen or of an amplicon generated from an appropriate NAAT assay, different from the NAAT assay initially used, should be obtained to provide a reliable test result [2].

In areas with spread of COVID-19 screening by rRT-PCR of a single Genomic target is considered sufficient. One or more negative results do not rule out the possibility of COVID-19 virus infection. A number of factors could lead to a negative result in an infected individual. These include poor quality of the specimen, containing little patient material, the specimen was collected late or very early in the infection, the specimen was not handled and shipped appropriately, technical reasons inherent in the test, e.g., virus mutation or PCR inhibition. If a negative result is obtained from a patient with a high index of suspicion for COVID-19 virus infection, particularly when only upper respiratory tract specimens were collected, additional specimens, including from the lower respiratory tract if possible, should be collected and tested [2].

Ensure that adequate standard operating procedures are in use and that staff are trained for appropriate specimen collection, storage, packaging and transport. All specimens collected for laboratory investigations should be regarded as potentially infectious. Ensure that health care workers who collect specimens adhere rigorously to infection prevention and control guidelines. Specific WHO interim guidance has been published [5, 6].

Antibody Testing

Several Laboratories in the world have now developed antibody testing for COVID-19 and rapid test kits are now commercially available, but the sensitivity is less when compared to PCR based tests [7]. Antibody tests are different because they are based on the knowledge of the proteins that form the viral coats, specifically, those proteins to which the immune system responds, triggering the production of antibodies that flag or neutralize the virus. Such viral protein coat must then be produced in the laboratory, using cell lines, for inclusion in an immunoassay (e.g., ELISA) that detects whether antibodies are present. Such immunoassays will form the basis of home testing kits for people who think they have had COVID-19 [8]. Antibody testing can be used as a screening tool in areas where there is clustering of cases.

General Lab Test Abnormalities in COVID19

Anemia was noted in about 15-25% of cases more commonly in non survivors [9]. Lymphocytopenia occurred in about 40%. Levels of d-dimer, high-sensitivity cardiac troponin I, serum ferritin, lactate dehydrogenase, and IL-6 were clearly elevated in non-survivors compared with survivors of COVID-19, throughout the clinical course, and increased duration of illness [9]. False positive dengue serology has been reported by some authors in COVID-19 [10].

Imaging in COVID19

X ray Chest

Chest X Ray may be normal in early stages. Later on bilateral multifocal consolidations can be seen. These may progress to involve entire lungs. Small pleural effusions may also develop in a few. Features of ARDS can be seen in critically ill cases

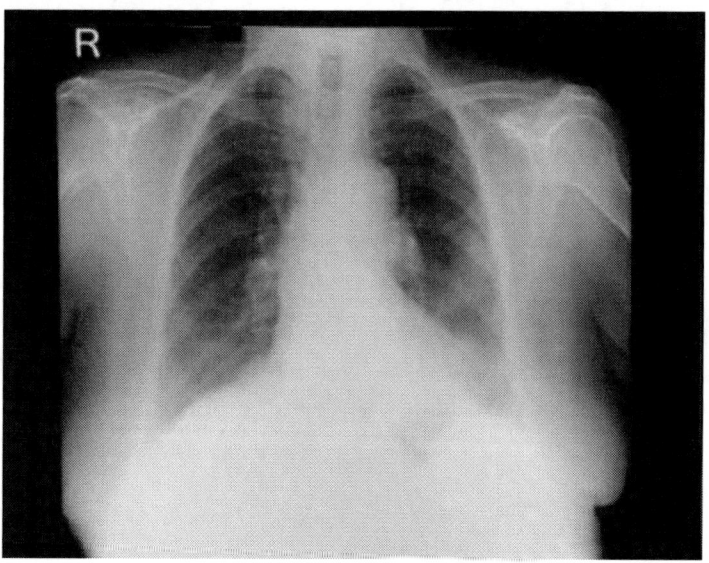

Figure 1. Xray chest-COVID-19.

CT Chest

CT scan can often be very helpful in diagnosis of COVID-19 [11]. Up to 50% of COVID-19 cases will have normal CT, 0-2 days after the onset of flu like symptom [11]. The findings that will help to suspect that COVID-19 infection are the ground are glass patterned areas, which, even in the initial stages, affect both lungs, in particular the lower lobes, and especially the posterior segments, with a fundamentally peripheral and subpleural distribution.

Findings in CT scan in Covid19 can be summarised as follows.

Early-up to 50-70% shows CT abnormalities early in the disease [11]. The main CT findings of COVID-19 infection are bilateral peripheral and basal predominant ground-glass opacity, consolidation, or both [6, 7]. Opacities often have an extensive distribution. Multiple discrete areas of ground-glass opacity, consolidation nor both occur in a subset of patients—often with round morphology or a reversed halo or atoll sign [12].

Pleural effusion, extensive tiny lung nodules, and lymphadenopathy occur in a very small number of cases and suggest bacterial super infection or another diagnosis [13].

Middle stage progression of lung opacities occur in middle stage of COVID-19. Peak lung involvement was characterized by development of crazy-paving (19%), new or increasing lung consolidation and higher rates of bilateral and multilobar involvement in up to86%.

Late-stage CT findings (14 days or longer) showed varying degrees of clearing but complete resolution may take at least 26 days [14].

REFERENCES

[1] World Health Organization. *Coronavirus disease* 2019 (COVID-19): situation report, 72. 2020.

[2] World Health Organization. *Laboratory testing for coronavirus disease 2019 (COVID-19) in suspected human cases: interim guidance*, 2 March 2020. World Health Organization; 2020.

[3] Zhang W, Du RH, Li B, Zheng XS, Yang XL, Hu B, et al. Molecular and serological investigation of 2019-nCoV infected patients: implication of multiple shedding routes. *Emerging Microbes & Infections* 2020;9(1):386-9.

[4] Zhang Y, Chen C, Zhu S, Shu C, Wang D, Song J, et al. Isolation of 2019-nCoV from a stool specimen of a laboratory-confirmed

case of the coronavirus disease 2019 (COVID-19). *China CDC Weekly* 2020;2(8):123-4.
[5] World Health Organization. Infection prevention and control during health care when COVID-19 is suspected: interim guidance, 19 March 2020. *World Health Organization*; 2020.
[6] World Health Organization. Laboratory biosafety guidance related to coronavirus disease 2019 (COVID-19): interim guidance, 12 February 2020. *World Health Organization*; 2020.
[7] Li Z, Yi Y, Luo X, Xiong N, Liu Y, Li S, et al. Development and Clinical Application of A Rapid IgMGÇÉIgG Combined Antibody Test for SARSGÇÉCoVGÇÉ2 Infection Diagnosis. *Journal of medical virology* 2020.
[8] Petherick A. Developing antibody tests for SARS-CoV-2. *The Lancet* 2020;395(10230):1101-2.
[9] Zhou F, Yu T, Du R, Fan G, Liu Y, Liu Z, et al. Clinical course and risk factors for mortality of adult inpatients with COVID-19 in Wuhan, China: a retrospective cohort study. *The Lancet* 2020.
[10] Yan G, Lee CK, Lam LT, Yan B, Chua YX, Lim AY, et al. Covert COVID-19 and false-positive dengue serology in Singapore. *The Lancet Infectious Diseases* 2020.
[11] Kanne JP, Little BP, Chung JH, Elicker BM, Ketai LH. *Essentials for radiologists on COVID-19: an update GÇöradiology scientific expert panel.* 2020. Radiological Society of North America.
[12] Pan F, Ye T, Sun P, Gui S, Liang B, Li L, et al. Time course of lung changes on chest CT during recovery from 2019 novel coronavirus (COVID-19) pneumonia. *Radiology* 2020;200370.
[13] Fang Y, Zhang H, Xie J, Lin M, Ying L, Pang P, et al. Sensitivity of chest CT for COVID-19: comparison to RT-PCR. *Radiology* 2020;200432.
[14] Bernheim A, Mei X, Huang M, Yang Y, Fayad ZA, Zhang N, et al. Chest CT findings in coronavirus disease-19 (COVID-19): relationship to duration of infection. *Radiology* 2020;200463.

Chapter 7

TRANSMISSION AND PREVENTION OF TRANSMISSION OF COVID-19

INTRODUCTION

SARS Cov2 virus which probably emerged originally from an animal source is now known to spread mainly through person-to-person. The virus generally spreads through respiratory droplets generated when a sick person coughs or sneezes or speaks. It may survive on surfaces that have been contaminated with respiratory secretions. Thus, contaminated surfaces may be another, less common; route of transmission. The virus can spread prior to the development of symptoms, in the incubation period. There have been confirmed cases of COVID-19 in people from asymptomatic patients and it acts as a potential threat in community transmission of the disease. Symptomatic persons are the most common source of transmission. Human to animal spread of the virus also has been reported.

Mode of Transmission

Droplet Transmission

The primary mode of spread is during close contact and by small droplets produced when patients cough, sneeze or talk; with close contact being within 1–3 meters [1].

A study in Hong Kong has confirmed that the virus was present in most patient's saliva in quantities reaching 100 million virus strands per 1 m [2]. A study found that an uncovered cough can lead to droplets travelling up to 4.5 meters [3]. A second study, during the current pandemic, found that droplets could travel even up to 8.2 meters (4).

Respiratory droplets are also produced while breathing out, and talking. The virus is not generally considered as airborne [5]. Indirect contact with surfaces in the immediate environment or with objects used on the infected person (e.g., stethoscope or thermometer) is another mode of transmission. People may also become infected by touching a contaminated surface and then their face [6]. The virus can survive on surfaces for up to 72 hours [7]. In COVID-19, airborne transmission may be possible in special circumstances in which procedures that generate aerosols are performed; i.e., end tracheal intubation, bronchoscopy, open suctioning, administration of nebulized treatment, manual ventilation before intubation, turning the patient to the prone position, disconnecting the patient from the ventilator, non-invasive positive-pressure ventilation, tracheostomy, and cardiopulmonary resuscitation [5]. The National Academy of Science has suggested that aerosol transmission may be possible and air collectors positioned in the hallway outside of people's rooms yielded samples positive for viral RNA [8]. The droplets can land in the mouths or noses of people who are nearby or possibly be inhaled into the lungs.

The virus is most contagious when people are symptomatic; while spread can occur possibly before the onset of symptoms, but this risk is low [9]. The virus survives for hours to days on surfaces [10]. The approximate period which the virus was found to be detectable were for one day on cardboard, for up to three days on plastic and stainless steel and for up to four hours on 99% copper [11]. This however may vary depending up on humidity and temperature.

Fecal-Oral Transmission

A possibility of feco-oral transmission is also considered for COVID-19 [12].

Studies of SARS indicated the gastrointestinal tract affinity for SARS coronavirus (SARS-CoV). The virus was detected in biopsy specimens and stool even in discharged patients, which may partially provide explanations for the gastrointestinal symptoms. ACE2 was not only highly expressed in the lung AT2 cells, but also in esophagus upper and stratified epithelial cells and absorptive enterocytes from ileum and colon. With the increasing gastrointestinal wall permeability to foreign pathogens once the virus infects, enteric symptoms like diarrhea will occur. Viral shedding from the digestive system might be longer-lasting than that from the respiratory tract. The findings suggest that we also need to use rectal swabs to confirm diagnosis of COVID-19 even after respiratory tract secretions become negative [12]. There are reports of mild to moderate liver injury including elevated aminotransferases, hypoproteinemia and prothrombin time prolongation in COVID-19.

Vertical transmission: Though several cases of delivery in covid19 patients has been reported, there are no convincing evidence at present for transplacental transmission [13].

Prevention of Transmission

Preventive strategies [14] remain the mainstay to control the spread of COVID-19. Performing hand hygiene frequently with an alcohol-based (more than 60%) hand rub if your hands are not visibly dirty or with soap and water if hands are dirty; avoiding touching your eyes, nose and mouth; practicing respiratory hygiene by coughing or sneezing into a bent elbow or tissue and then immediately disposing of the tissue; wearing a medical mask and performing hand hygiene after disposing of the mask, maintaining social distance (a minimum of 1 m) from individuals, social isolation of cases and contacts are the main measures employed. Wearing face mask when going to public places is an effective strategy which protect against spread of infection.

Health care professionals need to wear personal protective equipments (PPE) whenever they examine patients, or perform any procedures. All medical personnel involved in the management of patients with suspected COVID-19 must adhere to airborne precautions, hand hygiene, and donning of personal protective equipment. All aerosol-generating procedures should be done in an airborne infection isolation room. Double-gloving, might provide extra protection and minimize spreading via fomite contamination to the surrounding equipment after intubation. The intubation should be carried out by an expert, and avoid bag mask ventilation as far as possible.

Quarantine of Persons

Quarantine of persons is the restriction of activities of or the separation of persons who are not ill, but who may have been exposed to an infectious agent or disease [15]. This helps in early detection of cases and thus prevents its spread. Originally WHO recommends that contacts of patients with laboratory-confirmed COVID-19 be quarantined for 14 days from the last time they were exposed to the patient. The indications for quarantine include face-to-face contact with a COVID-19 patient within 1 meter and for >15 minutes; Providing direct care for patients with COVID-19 disease without using proper personal protective equipment, staying in the same close environment as a COVID-19 patient (including sharing a workplace, classroom or household or being at the same gathering) for any amount of time, travelling in close proximity with (that is, within 1 m separation from) a COVID-19 patient in any kind of conveyance [16].

MEASURES IN SUSPECTED AND CONFIRMED CASES

Suspected and confirmed patients should be isolated ideally in single rooms with bathroom facilities [17]. Patients should be educated regarding COVID-19 spread and proper hand washing and wearing of mask. Family visit should not be allowed but electronic communication should be encouraged for psychological support. Health care workers should be well educated about safety precautions and proper use of PPEs [18]. They may be divided in two teams and rotated with a period of quarantine in between. Proper personal hygiene, of health care workers is important, and avoiding contact with family members during duty period would be ideal to prevent spread of infection.

Disinfection

Floor and wall disinfection of COVID-19 isolation areas should be done with 1000mg/l chlorine containing disinfectant 3 times a day at least for 30 minutes [19]. Air sterilization can be carried out using a plasma sterilizer or ultraviolet lamp (1 hour) at least 3 times a day.

Fecal matter and sewage should be treated by treating with chlorine containing disinfectant with at least 1.5 hrs of disinfection time [19].

Blood/body fluid spills should be removed with disposable absorbent material soaked in 5000ml/l chlorine containing disinfectant solution. Will require higher concentration if spill volume is large [19]. Infectious fabrics, bed sheets, pillow covers etc. should be packed in 3 separate bags and should be sent to dedicated laundry. Wash with chlorine containing disinfectant at 90 degree at least for 30 minutes [19]. Medical waste should be collected in double layer bag separately labeled and send to disposal through specified path in specified time. Put sharp objects in special plastic container seal and spray with 1000mg/l chlorine containing disinfectant. Put the baggage waste into special medical waste transfer box with infection label [19].

Occupational Exposure –COVID-19

Intact/broken skin exposure –remove the contaminant, disinfectant to be applied at least for 3 minutes, either 75% alcohol or 0.5% iodophor, then flush with running water [19]. Exposure to mucus membrane - flush with saline or 0.055 iodophor. Sharp object injury- squeeze the blood if possible, wash the area with running water, disinfect with 755 alcohol/0.055 iodophor [19].

Direct exposure to respiratory tract - immediately leave the area, Gargle with plenty of normal saline or 0.05% iodophor, dip a cotton swab in 75% alcohol and clean the nasal cavity in a circular motion. Report to the relevant department and move to quarantine [19].

Handling of Dead Body

Standard precaution has to be followed by healthcare workers while handling the body including full PPE. 1% hypochlorate is used to for disinfection. Pack in double layer leak proof plastic body bags. All standard precautions should be taken at the time of burial. Disinfection of room elevators, hospital perm ices should be carried out with standard operating protocol [20].

Vaccine Development

The best way to prevent the infection is to develop an effective vaccine [21].

Efforts toward developing an effective SAR-CoV-2 vaccine have been initiated in many countries under WHO guidance. The target antigen selection and vaccine platform are probably based on SARS-CoV and MERS-CoV vaccine studies done previously. The genetic sequencing of SARSCov2 was made available in January 2020. A striking feature of the vaccine development for COVID-19 is the range of technology platforms being evaluated, including nucleic acid (DNA and RNA), virus-like particle, peptide, viral vector (replicating and non-replicating), recombinant protein, live attenuated virus and inactivated virus. All approaches are under phase-1 stage of development and will take a while to become commercially available.

REFERENCES

[1] Odega K, Iyamah E, Ibadin E, Idomeh F. *Safe Laboratory Practices in the Light of COVID-19 Pandemic: Way Forward in a Resource Limited Setting.* 2020.

[2] To KK-W, Tsang OT-Y, Yip CC-Y, Chan KH, Wu TC, Chan JM-C, et al. Consistent detection of 2019 novel coronavirus in saliva. *Clinical infectious diseases: an official publication of the Infectious Diseases Society of America 2020.*

[3] Loh NHW, Tan Y, Taculod J, Gorospe B, Teope AS, Somani J, et al. The impact of high-flow nasal cannula (HFNC) on coughing distance: implications on its use during the novel coronavirus disease outbreak. *Canadian Journal of Anesthesia/Journal canadien d'anesth+¬sie* 2020;1-2.

[4] Bourouiba L. Turbulent Gas Clouds and Respiratory Pathogen Emissions: Potential Implications for Reducing Transmission of COVID-19. *JAMA* 2020.

[5] World Health Organization. Modes of transmission of virus causing COVID-19: implications for IPC precaution recommendations: scientific brief, 27 March 2020. *World Health Organization*; 2020.

[6] Odega K, Iyamah E, Ibadin E, Idomeh F. *Safe Laboratory Practices in the Light of COVID-19 Pandemic: Way Forward in a Resource Limited Setting*. 2020.

[7] Salman FM, Abu-Naser SS, Alajrami E, Abu-Nasser BS, Alashqar BA. *COVID-19 Detection using Artificial Intelligence*. 2020.

[8] Loh NHW, Tan Y, Taculod J, Gorospe B, Teope AS, Somani J, et al. The impact of high-flow nasal cannula (HFNC) on coughing distance: implications on its use during the novel coronavirus disease outbreak. *Canadian Journal of Anesthesia/Journal canadien d'anesth+¬sie* 2020;1-2.

[9] Ammerman BA, Burke TA, Jacobucci R, McClure K. *Preliminary Investigation of the Association Between COVID-19 and Suicidal Thoughts and Behaviors in the US*. 2020.

[10] Bourouiba L. Turbulent gas clouds and respiratory pathogen emissions: potential implications for reducing transmission of COVID-19. *Jama 2020*.

[11] van Doremalen N, Bushmaker T, Morris DH, Holbrook MG, Gamble A, Williamson BN. & Lloyd-Smith, JO (2020). Aerosol and Surface Stability of SARS-CoV-2 as Compared with SARS-CoV-1. *New England Journal of Medicine*.

[12] Gu J, Han B, Wang J. COVID-19: Gastrointestinal Manifestations and Potential FecalGÇôOral Transmission. *Gastroenterology* 2020.

[13] Schwartz DA. An analysis of 38 pregnant women with COVID-19, their newborn infants, and maternal-fetal transmission of SARS-CoV-2: maternal coronavirus infections and pregnancy outcomes. *Archives of Pathology & Laboratory Medicine* 2020.

[14] World Health Organization. *Coronavirus disease 2019 (COVID-19): situation report*, 72. 2020.

[15] World Health Organization. Considerations for quarantine of individuals in the context of containment for coronavirus disease (COVID-19): interim guidance, 19 March 2020. *World Health Organization*; 2020.

[16] Sohrabi C, Alsafi Z, OGÇÖNeill N, Khan M, Kerwan A, Al-Jabir A, et al. World Health Organization declares global emergency: A review of the 2019 novel coronavirus (COVID-19*). International Journal of Surgery* 2020.

[17] Burke RM. Active monitoring of persons exposed to patients with confirmed COVID-19GÇöUnited States, JanuaryGÇôFebruary 2020. *MMWR Morbidity and mortality weekly report* 2020;69.

[18] World Health Organization. Rational use of personal protective equipment for coronavirus disease (COVID-19): interim guidance, 27 February 2020. *World Health Organization*; 2020.

[19] Liang T. *Handbook of COVID-19 prevention and treatment*. The First Affiliated Hospital, Zhejiang University School of Medicine Compiled According to Clinical Experience 2020.

[20] Ravi KS. Dead body management in times of COVID-19 and its potential impact on the availability of cadavers for medical education in India. *Anatomical Sciences Education* 2020.

[21] Lurie N, Saville M, Hatchett R, Halton J. Developing COVID-19 vaccines at pandemic speed. *New England Journal of Medicine* 2020.

Chapter 8

MANAGEMENT OF COVID-19

INTRODUCTION

COVID-19 spreads by person to person transmission and hence isolation of patient and proper personal protection by healthcare professionals is of paramount importance. Though so many trials are ongoing, at the time of writing this manuscript there is no definitive antiviral therapy or vaccine fully effective against COVID-19. The most commonly used option is a broad-spectrum antiviral drug like Nucleoside analogues and also HIV-protease inhibitors along with chloroquine. Some other drugs like azithromycin, ivermectin, plasmatherapy are used with varying success. In about 5% of patients critical care management will be required which will involve ventilator care treatment of ARDS, countering the cytokine storm and fluid management. High Mortality is associated with older age, comorbidities (including hypertension, diabetes, cardiovascular disease, chronic lung disease, and cancer), critical illness requiring ventilation, higher d-dimer and C-reactive protein concentrations, lower lymphocyte counts, and secondary infections.

SUPPORTIVE CARE

- Avoid using NSAID other than Paracetamol unless absolutely necessary
- Avoid using nebulized drug to prevent aerosolization of the virus, instead use metered dose inhaler.
- Antibiotic selection for secondary bacterial infection should be guided by the institutional antibiotic resistance pattern.
- non invasive ventilation should be avoided as there is risk of aerosol generation.
- Benefits and risks of steroids are unclear –so use judiciously

DRUGS FOR COVID-19

Chloroquine and Hydroxychloroquine

This drug was considered for covid from the observation that none of the lupus patient on this drug in Wuhan developed COVID-19 during the epidemic in Hubei. Several studies have tested the efficacy of chloloroquine in SARS infection and was found to be useful [1, 2]. There is sufficient pre-clinical data and evidence regarding the effectiveness of chloroquine for treatment of COVID-19 as well as evidence of safety from long-time use in other indications [3, 4, 5]. Chloroquine inhibited SARS-CoV-2 in vitro and suggested these drugs be assessed in human patients suffering from COVID-19 [6, 7]. Chloroquine slowed the progression of pneumonia and accelerated SARS-CoV-2 clearance and recovery in >100 patients with COVID-19, but results have not been published in the peer-reviewed literature and caution is advised in interpreting these findings [8]. With available evidence many physicians are using chloroquine for both prophylaxis [9, 10] and in the treatment of COVID-19. Experience (unpublished data) from kasarkode district of Kerala, India, 50 COVID-19 positive cases has been discharged without any mortality and all of them have received a combination of Chloroquine/hydroxychloroquine in combination with azithromycin (Dr Janardana Naik - Personal communication, Kazarkode Government Hospital, Kerala, India.)

Chloroquine prophylaxis dosage: Indian Council of Medical Research recommends use of prophylaxis in two situations [11].

1. Asymptomatic house hold contacts of COVID-19 confirmed cases
2. Asymptomatic health care workers

Asymptomatic health care workers should take 400mg of Hydroxychloroquine twice a day on day one, followed by once a week along with food for 7 weeks, asymptomatic contacts 400 mg twice a day followed by once a week for 3 weeks [11]. Drug is not recommended for persons with retinopathy, hypersensitivity or children below 15 yrs. Special care need to be taken in case of cardiac diseases and co administration with azithromycin as QT prolongation is a known side effect of the drug. In G6PD deficiency patients this drug can cause hemolysis. Acidic PH facilitates corona virus entry into the cell. HCQS alters the PH thus inhibiting the endocytosis. HCQS is more effective than chloroquine and reaches higher concentration in lung.

Though there are conflicting reports coming up regarding the efficacy of chloroquine in COVID-19, our experience with this drug along with azithromycin has been really good in Kerala. We had one of the lowest mortality of 0.53 as on 20th April 2020 which is one of the lowest reported in the world. There is a recent multicenter trial reporting chloroquine as ineffective in COVID-19 but our data from Kerala, India indicate definite benefit from the drug.

Lopinavir–Ritonavir

These are protease inhibitors. First drug tried in an RCT against COVID-19. RCT of lopinavir–ritonavir versus standard care in 199 hospitalized adults with SARS-CoV-2-associated pneumonia and hypoxemia but no benefit was observed in the controlled study [12]. Gastrointestinal side-effects, including diarrhea, nausea, and vomiting were noted with the drug [12]. Some trials showed benefit with these drugs [13, 14]. Recent recommendations by government of India do not advise this combination.

Azithromycin

A study with small sample size showed that hydroxychloroquine treatment significantly associated with viral load reduction/disappearance in COVID-19 patients and its effect is reinforced by azithromycin [15] [16]. *In vitro* effect of these combination was reported by some authors against SARS cov2 [17].

Oseltamivir

This drug used in the treatment of influenza has been used in several trials in COVID-19 but the results are not convincing enough to recommend the use [14, 18, 19].

Favipiravir (RNA-dependent RNA polymerase inhibitor)

Viral RNA polymerase inhibitor drug is best if given early. Several trials are underway with this drug in COVID-19, no definite recommendations can be made at this point of time [20, 21, 22].

Remdesivir (nucleotide analogue)

In vitro and *in vivo* trials has shown significant reduction in viral load on using this drug against SDARC cov2 virus [23, 24, 25, 26]. WHO has projected it as the most promising drug in COVID-19 and is currently being tested in 5 trials. Recently it has been approved by FDA based on the fact that it reduces the days of hospital admission in trials. As this drug acts by inhibiting viral replication, the chance of it to be effective is less because by the time the patient present in symptomatic state viral replication is already in peak. This drug reduces the total hospital stay when used in COVID-19 patients. It may be considered in patients with moderate disease (those on oxygen) with none of the following contraindications:

- AST/ALT > 5 times Upper limit of normal (ULN)
- Severe renal impairment (i.e., eGFR < 30ml/min/m^2 or need for hemodialysis)
- Pregnancy or lactating females
- Children (< 12 years of age)

Dose: 200 mg IV on day 1 followed by 100 mg IV daily for 4 days (total 5 days)

Tocilizumab (monoclonal antibody against interleukin-6)

Drug used in rheumatoid arthritis. Use in patients with elevated IL6 level. Cytokine storm has been attributed to the sudden deterioration of COVID-19 cases with ARDS. So antibody against interleukins was used in critically ill patients, but no convincing data available to recommend its use. Multi-centric trial is being done but results are not yet available. May be considered in patients with moderate disease with progressively increasing oxygen requirements and in mechanically ventilated patients not improving despite use of steroids. Long term safety data in COVID-19 remains largely unknown. Special considerations before its use include:

- Presence of raised inflammatory markers (e.g., CRP, Ferritin, IL-6)
- Patients should be carefully monitored post Tocilizumab for secondary infections and neutropenia
- The drug is contraindicated in PLHIV, those with active infections (systemic bacterial/fungal), Tuberculosis, active hepatitis, ANC < 2000/mm^3 and Platelet count < 1,00,000/mm^3

Sarilumab is another drug with similar anti IL6 activity.

Corticosteroids

An immune mediated mechanism is thought to be responsible for critical pulmonary involvement in COVID-19 and several patients has been tried with methyl prednisolone or other steroids, some trials do not give any evidence to support its use in COVID-19 [27, 28, 29]. But a recent trial has reported usefulness of dexamethasone. Consider IV methylprednisolone 0.5 to 1 mg/kg OR Dexamethasone 0.1 to 0.2 mg/kg for 3 days (preferably within 48 hours of admission or if oxygen requirement is increasing and if inflammatory markers are increased). Review the duration of administration as per clinical response.

Intravenous Immunoglobulin

Immunoglobulin was tried in some cases of COVID-19 but there is no convincing evidence to support its routine use [30, 31].

Convalescent Plasma

The use of plasma in critically ill COVID-19 infection has given promising results [32-34] however larger trials are required to recommend its use [35]. At present convalescent plasma can be considered in severe or critically ill COVID-19 patients and immuno suppressed patients. Some studies have given encouraging results with plasma and FDA has approved its use in COVID-19 [36]. Convalescent plasma may be considered in patients with moderate disease who are not improving (oxygen requirement is progressively increasing) despite use of steroids. Special prerequisites while considering convalescent plasma include:

- ABO compatibility and cross matching of the donor plasma
- Neutralizing titer of donor plasma should be above the specific threshold (if the latter is not available, plasma IgG titer (against S-protein RBD) above 1:640 should be used)
- Recipient should be closely monitored for several hours post transfusion for any transfusion related adverse events
- Use should be avoided in patients with IgA deficiency or immunoglobulin allergy

Dose: Dose is variable ranging from 4 to 13 ml/kg (usually 200 ml single dose given slowly over not less than 2 hours

Ivermectin

In vitro trials have given encouraging results with this drug but at present clinical datas to support the use are lacking [37].

On humanitarian basis several other drugs are being used in different parts of the world against COVID-19 which are beyond the scope of this chapter.

MANAGEMENT GUIDELINES

There are different guidelines followed in different hospitals, provinces and countries.

As a prototype the guidelines being followed in my state, Kerala of India is given below. As of today Kerala has one of the lowest mortality rates of 0.53% which is much lesser than the national average and mortality in other part of the world in cases of COVID-19 confirmed patients. Patients need to be categorized according to the clinical status as follows:

Categories

A	Mild sore throat/cough/rhinitis/diarrhea
B	Fever and/or severe sore throat/cough/diarrhea OR Category – A plus two or more of the following • Lung/heart/liver/kidney/neurological disease/hypertension/haematological disorders/uncontrolled diabetes/cancer/HIV-AIDS • On long term steroids/immunosuppressive drugs • Pregnant lady • Age more than 60 years OR • Category – A plus cardiovascular disease
C	• Breathlessness, chest pain, drowsiness, fall in blood pressure, haemoptysis, cyanosis (red flag signs) • Children with ILI (influenza like illness) with red flag signs (somnolence, high/persistent fever, inability to feed well, convulsions, dysnopea / respiratory distress etc. • Worsening of underlying chronic conditions

Identification of high risk patients is important and it is done based on the following features:

Identification of High Risk Patients

Co morbidities	clinical assessment	Lab values
Uncontrolled diabetes	Hypoxia Spo2 less than 93%	CPR more than 100mg/l
Hypertension	Tachycardia PR more than 125/mt	CPK more than twice upper limit
Cardiovascular disease	Respiratory rate more than 30/mt	
Lung disease	Hypotension BP90/60	Ferritin more than 300mcg/l
CKD	Alterered Sensorium	Trop t elevation
CLD		LDG more than 245U/L
On immunosuppressive		D Dimer more than 1000 ng/mL
HIV congenital immunodeficiency disorder		Multiorgan dysfunction
Age more than 60		ALC less than 0.8

Treatment strategies are decided based on the category.

MANAGEMENT OF MILD CASES

In the containment phase, patients with suspected or confirmed mild COVID-19 are being isolated to break the chain of transmission. Patients with mild disease may present to primary care/outpatient department, or detected during community outreach activities, such as home visits or by telemedicine.

Detailed clinical history is taken including that of co-morbidities. Patient is followed up daily for temperature, vitals and Oxygen saturation (SpO_2).

Patients should be monitored for signs and symptoms of complications that should prompt urgent referral. Patients with risk factors for severe illness should be monitored closely, given the possible risk of deterioration. If they develop any worsening symptoms (such as mental confusion, difficulty breathing, persistent pain or pressure in the chest, bluish coloration of face/lips, dehydration, decreased urine output, etc.), they should be immediately admitted to a Dedicated Covid Health Centre or Dedicated Covid Hospital.

Children with mild COVID-19 should be monitored for signs and symptoms of clinical deterioration requiring urgent re-evaluation. These include difficulty in breathing/fast or shallow breathing (for infants: grunting, inability to breastfeed), blue lips or face, chest pain or pressure, new confusion, inability to awaken/not interacting when awake, inability to drink or keep down any liquids.

Mild COVID-19 cases may be given:

1. Symptomatic treatment such as antipyretic (Paracetamol) for fever and pain, anti-tussives for cough
2. Adequate nutrition and appropriate hydration to ensured.
3. Tab Hydroxychloroquine (HCQ) may be considered for any of those having high risk features for severe disease (such as age> 60 years; Hypertension, diabetes, chronic lung/kidney/liver disease, Cerebrovascular disease and obesity) under strict medical supervision, preferably after shifting to DCHC/DCH.
4. Avoid HCQ in patients with underlying cardiac disease, history of unexplained syncope or QT prolongation (> 480 ms).

MANAGEMENT OF MODERATE CASES

Patients with suspected or confirmed moderate COVID-19 (pneumonia) is to be isolated to contain virus transmission. Patients with moderate disease may present to an emergency unit or primary care/outpatient department, or be encountered during community surveillance activities, such as active house to house search or by telemedicine.

The defining clinical assessment parameters are Respiratory Rate of more than or equal to 24 per minute and oxygen saturation (SpO$_2$) of less than 94% on room air (range 90-94%).

Such patients will be isolated in Dedicated Covid Health Centre.

The patient will undergo detailed clinical history including co-morbid conditions, measurement of vital signs, Oxygen saturation (SpO$_2$) and radiological examination of Chest X-ray, Complete Blood Count and other investigations as indicated.

Antibiotics should not be prescribed routinely unless there is clinical suspicion of a bacterial infection.

CLINICAL MANAGEMENT OF MODERATE CASES

1. Symptomatic treatment such as antipyretic (Paracetamol) for fever and pain, anti- tussives for cough
2. Adequate hydration to be ensured
3. Oxygen Support:
   - Target SpO$_2$: 92-96% (88-92% in patients with COPD)
   - The device for administering oxygen (nasal prongs, mask, or masks with breathing / non-rebreathing reservoir bag) depends upon the increasing requirement of oxygen therapy. If HFNC or simple nasal cannula is used, N95 mask should be applied over it.
   - Awake proning may be used as a rescue therapy.

Criteria to be fulfilled	Avoid proning
• Patients with oxygen requirement of >4L • Normal mental status • Able to self-prone or change position with minimal assistance	• Hemodynamic instability • Close monitoring not possible

Patients will undergo a rotational change in position from prone to lying on each side to sitting up. Typical protocols include 30–120 minutes in prone position, followed by 30– 120 minutes in left lateral decubitus, right lateral decubitus, and upright sitting position.

- All patients should have daily 12-lead ECG

1. Anticoagulation
   - Prophylactic dose of UFH or LMWH (e.g., enoxaparin 40 mg per day SC)

     *Contraindications: End stage renal disease, active bleeding, emergency surgery

     **Consider unfractionated heparin in ESRD

2. Corticosteroids
   - Consider IV methylprednisolone 0.5 to 1 mg/kg OR Dexamethasone 0.1 to 0.2 mg/kg for 3 days (preferably within 48 hours of admission or if oxygen requirement is increasing and if inflammatory markers are increased). Review the duration of administration as per clinical response.

3. Anti-virals
   - Tab. Hydroxychloroquine (400mg) BD on 1st day followed by 200mg 1 BD for 4 days. (after ECG Assessment)
   - May consider investigational therapies such as Remdesivir (under EUA); Convalescent.

4. Control of co-morbid condition

5. Follow up CRP, D-dimer & Ferritin every 48-72 hourly (if available); CBC with differential count, Absolute Lymphocyte count, KFT/LFT daily

6. Monitor for:
   - Increased work of breathing (use of accessory muscles)
   - Hemodynamic instability
   - Increase in oxygen requirement

If any of the above occurs, shift to Dedicated Covid Hospital

Few patients with COVID-19 experience a secondary bacterial infection. Consider empiric antibiotic therapy as per local antibiogram and guidelines in older people, immune- compromised patients, and children < 5 years of age.

Close monitoring of patients with moderate COVID-19 is required for signs or symptoms of disease progression. Provision of mechanisms for follow up and transportation to Dedicated Covid Hospital should be available.

MANAGEMENT OF SEVERE CASES

Early Supportive Therapy and Monitoring

1. Symptomatic treatment with paracetamol and antitussives to continue
2. Oxygenation: Give supplemental oxygen therapy immediately to patients with Severe Covid and respiratory distress, hypoxaemia, or shock: Initiate oxygen therapy at 5 L/min and titrate flow rates to reach target $SpO_2 \geq 90\%$ in non-pregnant adults and $SpO_2 \geq 92\text{-}96\%$ in pregnant patients. Children with emergency signs (obstructed or absent breathing, severe respiratory distress, central cyanosis, shock, coma or convulsions) should receive oxygen therapy during resuscitation to target $SpO_2 \geq 94\%$. All areas where patients with Severe Covid are cared for should be equipped with pulse oximeters, functioning oxygen systems and disposable, single- use, oxygen-delivering interfaces (nasal cannula, simple face mask, and mask with reservoir bag). Use contact precautions when handling contaminated oxygen interfaces of patients with COVID – 19.

3. Use conservative fluid management in patients with Severe Covid when there is no evidence of shock.
4. Anticoagulation: High prophylactic dose of UFH/ LMWH (e.g., enoxaparin 40 mg BD SC) if not at high risk of bleeding.
    *Contraindications: End stage renal disease, active bleeding, emergency surgery
    **Consider unfractionated heparin in ESRD
5. Corticosteroids: IV Methylprednisolone 1-2 mg/kg or Dexamethasone 0.2-0.4 mg/kg for 5-7 days
6. Investigational therapy: Tocilizumab (Off Label) Anti IL-6 therapy may be considered.

Management of Hypoxemic Respiratory Failure and ARDS

Recognize severe hypoxemic respiratory failure when a patient with respiratory distress is failing standard oxygen therapy. Patients may continue to have increased work of breathing or hypoxemia even when oxygen is delivered via a face mask with reservoir bag (flow rates of 10-15 L/min, which is typically the minimum flow required to maintain bag inflation; FiO2 0.60-0.95). Hypoxemic respiratory failure in ARDS commonly results from intrapulmonary ventilation- perfusion mismatch or shunt and usually requires mechanical ventilation.

Lung protective ventilation strategy by ARDS net protocol:

- Tidal volume 6 ml/kg, RR 15-35/min, PEEP 5-15cm H_2O; target plateau pressure < 30cm H_2O, target SpO_2 88-95% and/or PaO_2 55-80 mmHg

  Prone ventilation to be considered when there is refractory hypoxemia; PaO_2/FiO_2 ratio <150 with FiO_2> 0.6 with PEEP > 5 cm H_2O.

High – Flow Nasal Cannula oxygenation (HFNO) or non – invasive mechanical ventilation:

When respiratory distress and/or hypoxemia of the patient cannot be alleviated after receiving standard oxygen therapy, high – flow nasal cannula oxygen therapy or non – invasive ventilation can be considered. Compared to standard oxygen therapy, HFNO reduces the need for intubation. Patients with hypercapnia (exacerbation of obstructive lung disease, cardiogenic pulmonary oedema), hemodynamic instability, multi-organ failure, or abnormal mental status should generally not receive HFNO, although emerging data suggest that HFNO may be safe in patients with mild- moderate and non- worsening hypercapnia. Patients receiving HFNO should be in a monitored setting and cared for by experienced personnel capable of endotracheal intubation in case the patient acutely deteriorates or does not improve after a short trial (about 1 hr).

NIV: setting - PS 5-15 cmH2O adjusted to tidal volume of 5-7 ml/kg and PEEP 5-10 cm H2O and FiO_2 @ 0.5 -1.0 titrated to target SpO_2 > 94%.

There have been concerns raised about generation of aerosols while using HFNO and NIV. However, recent publications suggest that newer HFNO and NIV systems with good interface fitting do not create widespread dispersion of exhaled air and therefore should be associated with low risk of airborne transmission. If conditions do not improve or even get worse within a short time (1–2 hours), tracheal intubation and invasive mechanical ventilation should be used in a timely manner.

- Endotracheal intubation should be performed by a trained and experienced provider using airborne precautions. Patients with ARDS, especially young children or those who are obese or pregnant, may de-saturate quickly during intubation. Pre-oxygenate with 100% FiO2 for 5 minutes, via a face mask with reservoir bag, bag-valve mask, HFNO, or NIV. Rapid sequence intubation is appropriate after an airway assessment that identifies no signs of difficult intubation.
- Implement mechanical ventilation using lower tidal volumes (4–8 ml/kg predicted body weight, PBW) and lower inspiratory pressures (plateau pressure <30 cm H_2O). This is a strong recommendation from a clinical guideline for patients with ARDS, and is suggested for patients with sepsis-induced respiratory failure. The initial tidal volume is 6 ml/kg PBW; tidal volume up to 8 ml/kg PBW is allowed if undesirable side effects occur (e.g., dys-synchrony, pH < 7.15). Hypercapnia is permitted if meeting the pH goal of 7.30-7.45. Ventilator protocols are available. The use of deep sedation may be required to control respiratory drive and achieve tidal volume targets.
- In patients with severe ARDS, prone ventilation for 16-18 hours per day is recommended but requires sufficient human resources and expertise to be performed safely.
- In patients with moderate or severe ARDS, higher PEEP instead of lower PEEP is suggested. PEEP titration requires consideration of benefits (reducing atelect trauma and improving alveolar recruitment) vs. risks (end-inspiratory over distension leading to lung injury and higher pulmonary vascular resistance). Tables are available to guide PEEP titration based on the FiO_2 required to maintain SpO_2. In patients with moderate- severe ARDS ($PaO_2/FiO_2<150$), neuromuscular blockade by continuous infusion should not be routinely used.

Management of COVID-19

- In settings with access to expertise in extracorporeal life support (ECLS), consider referral of patients with refractory hypoxemia despite lung protective ventilation. ECLS should only be offered in expert centres with a sufficient case volume to maintain expertise and that can apply the IPC measures required for COVID – 19 patients.
- Avoid disconnecting the patient from the ventilator, which results in loss of PEEP and atelectasis. Use in-line catheters for airway suctioning and clamp endotracheal tube when disconnection is required (for example, transfer to a transport ventilator).

Management of Septic Shock

- Recognize septic shock in adults when infection is suspected or confirmed AND vasopressors are needed to maintain mean arterial pressure (MAP) ≥65 mmHg AND lactate is >2 mmol/L, in absence of hypovolemia. Recognize septic shock in children with any hypotension (systolic blood pressure [SBP] <5th centile or >2 SD below normal for age) or two of the three of the following: altered mental state; tachycardia or bradycardia (HR <90 bpm or >160 bpm in infants and HR<70 bpm or >150 bpm in children); prolonged capillary refill (>2 sec) or warm vasodilation with bounding pulses; tachypnea; mottled skin or petechial or purpuric rash; increased lactate; oliguria; hyperthermia or hypothermia.
- In the absence of a lactate measurement, use MAP and clinical signs of perfusion to define shock. Standard care includes early recognition and the following treatments within 1 hour of recognition: antimicrobial therapy and fluid loading and vasopressors for hypotension. The use of central venous and

arterial catheters should be based on resource availability and individual patient needs.
- In resuscitation from septic shock in adults, give at least 30 ml/kg of isotonic crystalloid in adults in the first 3 hours. In resuscitation from septic shock in children in well-resourced settings, give 20 ml/kg as a rapid bolus and up to 40-60 ml/kg in the first 1 hr. Do not use hypotonic crystalloids, starches, or gelatins for resuscitation.
- Fluid resuscitation may lead to volume overload, including respiratory failure. If there is no response to fluid loading and signs of volume overload appear (for example, jugular venous distension, crackles on lung auscultation, pulmonary oedema on imaging, or hepatomegaly in children), then reduce or discontinue fluid administration. This step is particularly important where mechanical ventilation is not available. Alternate fluid regimens are suggested when caring for children in resource-limited settings.
- Crystalloids include normal saline and Ringer's lactate. Determine need for additional fluid boluses (250-1000 ml in adults or 10-20 ml/kg in children) based on clinical response and improvement of perfusion targets. Perfusion targets include MAP (>65 mmHg or age-appropriate targets in children), urine output (>0.5 ml/kg/hr in adults, 1 ml/kg/hr. in children), and improvement of skin mottling, capillary refill, level of consciousness, and lactate. Consider dynamic indices of volume responsiveness to guide volume administration beyond initial resuscitation based on local resources and experience. These indices include passive leg raising test, fluid challenges with serial stroke volume measurements, or variations in systolic pressure, pulse pressure, inferior vena cava size, or stroke volume in response to changes in intrathoracic pressure during mechanical ventilation.

- Administer vasopressors when shock persists during or after fluid resuscitation. The initial blood pressure target is MAP ≥ 65 mmHg in adults and age-appropriate targets in children.
- If central venous catheters are not available, vasopressors can be given through a peripheral IV, but use a large vein and closely monitor for signs of extravasation and local tissue necrosis. If extravasation occurs, stop infusion. Vasopressors can also be administered through intraosseous needles.
- If signs of poor perfusion and cardiac dysfunction persist despite achieving MAP target with fluids and vasopressors, consider an inotrope such as dobutamine.

OTHER THERAPEUTIC MEASURES

For patients with progressive deterioration of oxygenation indicators, rapid worsening on imaging and excessive activation of the body's inflammatory response, glucocorticoids can be used for a short period of time (3 to 5 days). It is recommended that dose should not exceed the equivalent of Methylprednisolone 1 – 2mg/kg/day OR Dexamethasone 0.2-0.4 mg/kg/day. Note that a larger dose of glucocorticoid will delay the removal of coronavirus due to immunosuppressive effects.

Prophylactic dose of UFH or LMWH (e.g., enoxaparin 40 mg per day SC) should be given for anti-coagulation. Control of co-morbid conditions should be ensured.

For pregnant severe cases, consultations with obstetric, neonatal, and intensive care specialists (depending on the condition of the mother) are essential. Patients often suffer from anxiety and fear and they should be supported by psychological counseling.

For patients with evidence of cytokine release Syndrome

Grade	Clinical assessment	Treatment
Grade-1	Mild reaction, low grade Fever, No oxygen requirement	No Treatment
Grade 2	Moderate reaction High grade fever(more than 103F, need for IVF(No Hypotension, mild oxygen requirement less than 6l/mt Grade 2 AKI Grade 3 LFT(Raised liver enzymes and s. Bilirubin above 2.5gm/dl	Send for serum IL6 If not available use CRP as a surrogate marker.
Grade 3	Severe Reaction Rapidly worsening respiratory status with radiographic infiltrates and Spo2 less than 93% in room air or on supplemental oxygen(more than 6l/mt high flow, BiPAP/CPAP) Grade4 liverfunction test (raised liver enzymes, S. Bilirubin more than 2.5 mg/dl, IMD more than 1.5, encephalopathy) Grade3 AKI, IVF for resuscitation, Coagulopathy requiringcorrection with FFP or cryoprecipitate Low dose vasopressor(Noradrenaline 0.5 mcg/Kg/mt	Send for Serum IL-6 or CRP Consider tocilizumab More than 18 yrs 8mg/kg IV max 400 mg
Grade 4	Life threatening multiorgan dysfunction, hypoxia requiring mechanical ventilation, Hypotension requiring high dose vasopressors	Send for Serum IL-6 or CRP Consider tocilizumab

WHO -Case definition of Clinical Syndromes [38]

Mild illness	Patients with uncomplicated upper respiratory tract viral infection, may have non-specific symptoms such as fever, fatigue, cough (with or without sputum production), anorexia, malaise, muscle pain, sore throat, dyspnea, nasal congestion, or headache. Rarely, patients may also present with diarrhoea, nausea and vomiting. The elderly and immunosuppressed may present with atypical symptoms. Symptoms due to physiologic adaptations of pregnancy or adverse pregnancy events, such as e.g. dyspnea, fever, GI-symptoms or fatigue, may overlap with COVID-19
Pneumonia	Adult with pneumonia but no signs of severe pneumonia and no need for supplemental oxygen. Child with non-severe pneumonia who has cough or difficulty breathing + fast breathing: fast breathing (in breaths/min): < 2 months: $\geq$ 60; 2–11 months: $\geq$ 50; 1–5 years: $\geq$ 40, and no signs of severe pneumonia

Severe Pneumonia	Adolescent or adult: fever or suspected respiratory infection, plus one of: respiratory rate > 30 breaths/min; severe respiratory distress; or SpO2 ≤ 93% on room air.
	Child with cough or difficulty in breathing, plus at least one of the following: central cyanosis or SpO_2 < 90%; severe respiratory distress (e.g. grunting, very severe chest indrawing); signs of pneumonia with a general danger sign: inability to breastfeed or drink, lethargy or unconsciousness, or convulsions (15). Other signs of pneumonia may be present: chest in drawing, fast breathing (in breaths/min): < 2 months: ≥ 60; 2–11 months: ≥ 50; 1–5 years: ≥ 40 (16). While the diagnosis is made on clinical grounds; chest imaging may identify or exclude some pulmonary complications.
ARDS	Onset: within 1 week of a known clinical insult or new or worsening respiratory symptoms.
	Chest imaging (radiograph, CT scan, or lung ultrasound): bilateral opacities, not fully explained by volume overload, lobar or lung collapse, or nodules.
	Origin of pulmonary infiltrates: respiratory failure not fully explained by cardiac failure or fluid overload. Need objective assessment
	(e.g., echocardiography) to exclude hydrostatic cause of infiltrates/oedema if no risk factor present.
	Oxygenation impairment in adults (17, 19):
	Mild ARDS: 200 mmHg < PaO_2/FiO_2a ≤ 300 mmHg (with PEEP or CPAP ≥ 5 cm H_2O, or non-ventilated)
	Moderate ARDS: 100 mmHg < PaO_2/FiO_2 ≤ 200 mmHg (with PEEP ≥ 5 cmH2O, or non-ventilated)
	Severe ARDS: PaO_2/FiO_2 ≤ 100 mmHg (with PEEP ≥ 5 cm H_2O, or non-ventilated)
	When PaO_2 is not available, SpO_2/FiO_2 ≤ 315 suggests ARDS (including in non-ventilated patients).
	Oxygenation impairment in children: note OI = Oxygenation Index and OSI = Oxygenation Index using SpO_2. Use PaO_2-based metric when available. If PaO_2 not available, wean FiO_2 to maintain SpO_2 ≤ 97% to calculate OSI or SpO_2/FiO_2 ratio:
	Bilevel (NIV or CPAP) ≥ 5 cm H_2O via full face mask: PaO_2/FiO_2 ≤ 300 mmHg or SpO_2/FiO_2 ≤ 264
	Mild ARDS (invasively ventilated): 4 ≤ OI < 8 or 5 ≤ OSI < 7.5
	Moderate ARDS (invasively ventilated): 8 ≤ OI < 16 or 7.5 ≤ OSI < 12.3
	Severe ARDS (invasively ventilated): OI ≥ 16 or OSI ≥ 12.3.
Sepsis	Adults: life-threatening organ dysfunction caused by a dysregulated host response to suspected or proven infection. Signs of organ dysfunction include: altered mental status, difficult or fast breathing, low oxygen saturation, reduced urine output, fast heart rate, weak pulse, cold extremities or low blood pressure, skin mottling, or laboratory evidence of coagulopathy,
	thrombocytopenia, acidosis, high lactate or hyperbilirubinemia.
	Children: suspected or proven infection and ≥ 2 aged based systemic inflammatory response syndrome criteria, of which one must be abnormal temperature or white blood cell count

(Continued)

Septic Shock	Adults: persisting hypotension despite volume resuscitation, requiring vasopressors to maintain MAP ≥ 65 mmHg and serum lactate level > 2 mmol/L. Children: any hypotension (SBP < 5th centile or > 2 SD below normal for age) or two or three of the following: altered mental state; tachycardia or bradycardia (HR < 90 bpm or > 160 bpm in infants and HR < 70 bpm or > 150 bpm in children); prolonged capillary refill (> 2 sec) or feeble pulse; tachypnea; mottled or cool skin or petechial or purpuric rash; increased lactate; oliguria; hyperthermia or hypothermia.

OTHER INTENSIVE MANAGEMENTS-CRITICALLY ILL

- Proper nutritional support
- Renal replacement therapy if needed;
- early physical therapy;
- prevention of nosocomial infections;
- deep vein thrombosis prophylaxis;
- stress ulcer prophylaxis [41]
- Rehabilitation therapy
- Early self-proning in awake, non-intubated patients
- Any COVID-19 patient with respiratory embarrassment severe enough to be admitted to the hospital may be considered for rotation and early self-proning.
- Care must be taken to not disrupt the flow of oxygen during patient rotation
- Typical protocols include 30–120 minutes in prone position, followed by 30–120 minutes in left lateral decubitus, right lateral decubitus, and upright sitting position

REQUIREMENTS FOR SAFE PRONE POSITIONING IN ARDS

Pre-oxygenate the patient with FiO2 1.0
Secure the endotracheal tube and arterial and central venous catheters
Adequate number of staff to assist in the turn and to monitor the turn
Supplies to turn (pads for bed, sheet, protection for the patient)
Knowledge of how to perform the turn as well as how to supine the patient in case of an emergency
Contraindications to prone ventilation
Spinal instability requires special care
Intra cranial pressure may increase on turning
Rapidly return to supine in case of CPR or defibrillation
When to start proning?
P/F ratio <150 while being ventilated with FiO2 >0.6 and PEEP >5 cm H2O When to stop proning?
When P/F exceeds 150 on FiO2 > 0.6 and > 6 PEEP
What portion of the day should patients be kept prone?
As much as possible (16-18 hours a day)
Adult patients with severe ARDS receive prone positioning for more than 12 hours per day.

DISCHARGE PROTOCOL

Discharge is considered when

- Body temperature remains normal for at least 3 days.
- PCR test is negative
- Significant improvement in respiratory symptoms
- Lung imaging features improved

- No complications
- After discharge patient should be in home isolation at least two weeks

Conclusion

The above recommendations are largely based on Indian recommendations (ICMR) in COVID-19 and the recommendations will vary from country to country.

Management of COVID-19 is also fast evolving; Hence readers are advised to check the latest guidelines when managing COVID-19 cases.

References

[1] Rolain JM, Colson P, Raoult D. Recycling of chloroquine and its hydroxyl analogue to face bacterial, fungal and viral infections in the 21st century. *International journal of antimicrobial agents* 2007;30(4):297-308.

[2] Keyaerts E, Vijgen L, Maes P, Neyts J, Van Ranst M. In vitro inhibition of severe acute respiratory syndrome coronavirus by chloroquine. *Biochemical and biophysical research communications* 2004;323(1):264-8.

[3] Cortegiani A, Ingoglia G, Ippolito M, Giarratano A, Einav S. A systematic review on the efficacy and safety of chloroquine for the treatment of COVID-19. *Journal of critical care* 2020.

[4] Devaux CA, Rolain JM, Colson P, Raoult D. New insights on the antiviral effects of chloroquine against coronavirus: what to expect for COVID-19? *International journal of antimicrobial agents* 2020;105938.

[5] Gao J, Tian Z, Yang X. Breakthrough: Chloroquine phosphate has shown apparent efficacy in treatment of COVID-19 associated pneumonia in clinical studies. *Bioscience trends* 2020.

[6] Wang L. *Cell Res*. 2019. the press.

[7] Gautret P, Lagier JC, Parola P, Meddeb L, Mailhe M, Doudier B, et al. Hydroxychloroquine and azithromycin as a treatment of COVID-19: results of an open-label non-randomized clinical trial. *International journal of antimicrobial agents* 2020;105949.

[8] Gao J, Tian Z, Yang X. Breakthrough: Chloroquine phosphate has shown apparent efficacy in treatment of COVID-19 associated pneumonia in clinical studies. *Bioscience trends* 2020.

[9] Pourdowlat G, Panahi P, Pooransari P, Ghorbani F. Prophylactic Recommendation for Healthcare Workers in COVID-19 Pandemic. *Advanced Journal of Emergency Medicine* 2020.

[10] Mor N. *Resources for primary care providers to meet patients needs during the COVID-19 epidemic*. 2020.

[11] Agrawal S, Goel AD, Gupta N. Emerging prophylaxis strategies against COVID-19. *Monaldi Archives for Chest Disease* 2020;90(1).

[12] Cao B, Wang Y, Wen D, Liu W, Wang J, Fan G, et al. A trial of lopinavirGÇôritonavir in adults hospitalized with severe Covid-19. *New England Journal of Medicine* 2020.

[13] Velavan TP, Meyer CG. The COVID-19 epidemic. *Trop Med Int Health* 2020;25(3):278-80.

[14] Velavan TP, Meyer CG. The COVID-19 epidemic. *Trop Med Int Health* 2020;25(3):278-80.

[15] Gautret P, Lagier JC, Parola P, Meddeb L, Mailhe M, Doudier B, et al. Hydroxychloroquine and azithromycin as a treatment of COVID-19: results of an open-label non-randomized clinical trial. *International journal of antimicrobial agents* 2020;105949.

[16] Dahly D, Gates S, Morris T. *Statistical review of Hydroxychloroquine and azithromycin as a treatment of COVID-*

19: results of an open-label non-randomized clinical trial. (Version 1.0). Zenodo. Preprint Posted online 2020;23.

[17] Andreania J, Le Bideaua M, Duflota I, Jardota P, Rollanda C, Boxbergera M, et al. In vitro testing of Hydroxychloroquine and Azithromycin on SARS-CoV-2 shows 1 synergistic effect 2. *lung* 2020;21:22.

[18] Wu Y, Xie Yl, Wang X. Longitudinal CT findings in COVID-19 pneumonia: Case presenting organizing pneumonia pattern. Radiology: *Cardiothoracic Imaging* 2020;2(1):e200031.

[19] Sohrabi C, Alsafi Z, OGÇÖNeill N, Khan M, Kerwan A, Al-Jabir A, et al. World Health Organization declares global emergency: A review of the 2019 novel coronavirus (COVID-19). *International Journal of Surgery* 2020.

[20] Cai Q, Yang M, Liu D, Chen J, Shu D, Xia J, et al. Experimental treatment with favipiravir for COVID-19: an open-label control study. *Engineering* 2020.

[21] Dong L, Hu S, Gao J. Discovering drugs to treat coronavirus disease 2019 (COVID-19). *Drug discoveries & therapeutics* 2020;14(1):58-60.

[22] Chang YC, Tung YA, Lee KH, Chen TF, Hsiao YC, Chang HC, et al. *Potential therapeutic agents for COVID-19 based on the analysis of protease and RNA polymerase docking.* 2020.

[23] Wang M, Cao R, Zhang L, Yang X, Liu J, Xu M, et al. Remdesivir and chloroquine effectively inhibit the recently emerged novel coronavirus (2019-nCoV) *in vitro. Cell research* 2020;30(3):269-71.

[24] Ko WC, Rolain JM, Lee NY, Chen PL, Huang CT, Lee PI, et al. Arguments in favor of remdesivir for treating SARS-CoV-2 infections. *Int J Antimicrob Agents* 2020;105933.

[25] Zhang L, Zhou R. *Binding Mechanism of Remdesivir to SARS-CoV-2 RNA Dependent RNA Polymerase.* 2020.

[26] Al-Tawfiq JA, Al-Homoud AH, Memish ZA. Remdesivir as a possible therapeutic option for the COVID-19. *Travel Med Infect Dis* 2020;101615.

[27] Mehta P, McAuley DF, Brown M, Sanchez E, Tattersall RS, Manson JJ. COVID-19: consider cytokine storm syndromes and immunosuppression. *The Lancet* 2020.

[28] Singhal T. A review of coronavirus disease-2019 (COVID-19). *The Indian Journal of Pediatrics* 2020;1-6.

[29] Wang Y, Jiang W, He Q, Wang C, Wang B, Zhou P, et al. Early, low-dose and short-term application of corticosteroid treatment in patients with severe COVID-19 pneumonia: single-center experience from Wuhan, China. *medRxiv* 2020.

[30] Zhu L, Xu X, Ma K, Yang J, Guan H, Chen S, et al. Successful recovery of COVIDGÇÉ19 pneumonia in a renal transplant recipient with longGÇÉterm immunosuppression. *American Journal of Transplantation* 2020.

[31] Wu J, Liu J, Zhao X, Liu C, Wang W, Wang D, et al. Clinical characteristics of imported cases of COVID-19 in Jiangsu province: a multicenter descriptive study. *Clinical infectious diseases: an official publication of the Infectious Diseases Society of America 2020.*

[32] Shen C, Wang Z, Zhao F, Yang Y, Li J, Yuan J, et al. Treatment of 5 critically ill patients with COVID-19 with convalescent plasma. *JAMA* 2020.

[33] Casadevall A, Pirofski La. The convalescent sera option for containing COVID-19. *The Journal of clinical investigation* 2020;130(4).

[34] Chen L, Xiong J, Bao L, Shi Y. Convalescent plasma as a potential therapy for COVID-19. *The Lancet Infectious Diseases* 2020;20(4):398-400.

[35] Shen C, Wang Z, Zhao F, Yang Y, Li J, Yuan J, et al. Treatment of 5 critically ill patients with COVID-19 with convalescent plasma. *JAMA* 2020.

[36] Tanne JH. Covid-19: FDA approves use of convalescent plasma to treat critically ill patients. 2020. *British Medical Journal Publishing Group*.

[37] Caly L, Druce JD, Catton MG, Jans DA, Wagstaff KM. The FDA-approved Drug Ivermectin inhibits the replication of SARS-CoV-2 *in vitro*. *Antiviral Research* 2020;104787.

[38] World Health Organization. Clinical management of severe acute respiratory infection (SARI) when COVID-19 disease is suspected: interim guidance, 13 March 2020. *World Health Organization*; 2020.

[39] Cascella M, Rajnik M, Cuomo A, Dulebohn SC, Di Napoli R. Features, evaluation and treatment coronavirus (COVID-19). *StatPearls* [Internet]. StatPearls Publishing; 2020.

[40] MacLaren G, Fisher D, Brodie D. Preparing for the most critically ill patients with COVID-19: the potential role of extracorporeal membrane oxygenation. *JAMA* 2020.

[41] Murthy S, Gomersall CD, Fowler RA. Care for critically ill patients with COVID-19. *JAMA* 2020.

Chapter 9

PROGNOSIS OF COVID-19

INTRODUCTION

Prognosis of COVID-19 is generally good provided the patients are not in extremes of age or have associated Comorbidities. The mortality rate reported varies from country to country and depends on availability of best medical care for critically ill patients. About 5% of COVID-19 patients become critically ill requiring intensive care. When huge number of people is infected with SARC cov2 giving best care to critically ill patients become an issue even for medically advanced countries. Hence the best way to reduce the death rate is to reduce the number of people getting infected with the virus. As no vaccine is available at present this can be best achieved by lock downs, social distancing and isolation of cases and quarantine of exposed individuals with potential risk of developing the infection.

The reported case fatality rates in COVID-19 vary from 1% to more than 7%. These values must be interpreted with caution. For example, in countries where massive screening has been performed in the whole population (e.g., in South Korea and Switzerland), overall case fatality rates of less than 1% have been reported, because the denominator included many mild or asymptomatic cases. However, in countries where only people requiring hospital admission were screened (e.g., Italy and Spain), very high mortality is reported, exceeding 5%, because the denominator is much smaller.

The factors associated with high mortality include: age $\geq$ 65 years, hypertension, cardiovascular or cerebro vascular diseases, dyspnea, fatigue, sputum production, headache. Investigation abnormalities like white blood cell counts > 10× 109/L, neutrophils > 6.3 × 109/L, CD3+CD8+ T cells $\leq$ 75 cell/μL, cardiac troponin I $\geq$ 0.05 ng/mL, myoglobin > 100 ng/L, creatinine $\geq$ 133 μmol/L, D-dimer $\geq$ 0.5 mg/L, and BP < 60 mmHg were associated with high death rates in patients with COVID-19 pneumonia [2].

CD3+CD8+ T cells $\leq$ 75 cell/μL, was a reliable predictor for mortality of patient with COVID-19 pneumonia [2]. Progressive immune-associated injury and inadequate adaptive immune responses were possible mechanisms by which SARS-CoV-2 causes severe illness and high mortality [2].

On analysis of Italian data the case-fatality rate in Italy and China appear very similar for age groups 0 to 69 years, but rates were higher in Italy among individuals aged 70 years or older, and in particular among those aged 80 years or older [3]. In age group above 90 have a very high fatality rate (22.7%). Reports from China shows fatality rate among patients aged 80 years or older that was similar to the rate in the Italian sample (21.9% in China vs 20.2% in Italy [3]. In the elderly there were several co-existing diseases that was contributing to the high mortality [3]. A subgroup analysis of Italian patient who died showed mean age of death was 79.5 years. Out of them 30% had ischemic heart disease, 35.5% had diabetes, 20.3% had active cancer, 24.5% had atrial fibrillation, 6.8% had dementia, and 9.6% had a history of stroke. The mean number of pre-existing diseases was 2.7 (SD, 1.6). Overall 0.8% had no diseases, 25.1% had a single disease, 25.6% had 2 diseases and 48.5% had 3 or more underlying diseases [3].

In short, predictors of a fatal outcome in COVID-19 cases include old age, the presence of underlying diseases, the presence of secondary infection and elevated inflammatory indicators in the blood. In critical cases COVID-19 mortality might be due to virus-activated "cytokine storm syndrome" or fulminant myocarditis [4].

REFERENCES

[1] Chen N, Zhou M, Dong X, Qu J, Gong F, Han Y, et al. Epidemiological and clinical characteristics of 99 cases of 2019 novel coronavirus pneumonia in Wuhan, China: a descriptive study. *The Lancet* 2020;395(10223):507-13.

[2] Du RH, Liang LR, Yang CQ, Wang W, Cao TZ, Li M, et al. Predictors of Mortality for Patients with COVID-19 Pneumonia Caused by SARS-CoV-2: A Prospective Cohort Study. *European Respiratory Journal* 2020.

[3] Onder G, Rezza G, Brusaferro S. Case-fatality rate and characteristics of patients dying in relation to COVID-19 in Italy. *JAMA* 2020.

[4] Ruan Q, Yang K, Wang W, Jiang L, Song J. Clinical predictors of mortality due to COVID-19 based on an analysis of data of 150 patients from Wuhan, China. *Intensive care medicine* 2020;1-3.

Chapter 10

KERALA HEALTH CARE SYSTEM AND COVID-19

INTRODUCTION

Preparing the health care system plays the key role in preventing the spread of COVID-19 pandemic. Government's decisions are mostly based on previous experiences and models from other countries. Measures are guided by WHO recommendations, infrastructural availability and availability of health care professionals. Kerala model, which they adopted in the Kerala state, of India is a typical example of successful health administration implemented through public health centers with focus on contact identification and home quarantine. Symptomatic contacts were treated in special covid care hospitals with dedicated staff, so that there was no nosocomial spread of infection. First case of COVID-19 in India was reported on 30th January 2020 in Kerala in a student returned from Wuhan [1]. Kerala which has a population of 36 million has total of 394 covid cases as on 16[h] April 2020 and a total of 2 deaths. In short Kerala has effectively controlled the community spread by aggressive tracking, testing and social isolation and by implementation of social distancing. With the lowest mortality, the Kerala model of covid care has been appreciated throughout the country and abroad.

HIGHLIGHTS OF KERALA HEALTHCARE MODEL AND STRATEGIES FOR PREVENTION OF SPREAD OF COVID-19 [2]

Kerala has lowest death rate amongst the states of India due to COVID-19. Out of 394 cases reported as on 16th April 2020, 245 has recovered and only 2 deaths were reported. The government of Kerala declared high alert from 4th February. Early February saw the first wave of symptomatic people in the state, with nearly 3,500 people put under isolation [3]. All people returning from abroad were screened at the airport, and those with symptoms were admitted. Those without symptoms were advised 14 days home quarantine. Swabs were taken for PCR in all symptomatic cases. All PCR positive cases were subjected to strict route mapping. Those who have come into contact were advised home quarantine and to report to the government hospital if they develop any covid symptoms. Asha workers and health inspectors who were primary health care providers in the community were asked to keep strict vigil on quarantined people by regular visit. Isolation wards with 40 beds were set up in 21 major hospitals of the state and a helpline was activated in every District, Kerala has 14 districts. As of April 8, 2020, Kerala had over 140 thousand contacts under isolation. Of these, around 139 thousand people were confined to their homes, while 749 patients were quarantined in hospitals. As of 4 March, 215 health care workers were deployed across Kerala and 3,646 tele - counseling services were conducted to provide psycho-social support to families of those suspected to be infected and the general public regarding the status of corona virus spread in Kerala and precautions to be taken. There are three corona virus testing centers in

Kerala: National Institute of Virology Field Unit, Thiruvananthapuram Medical College and Calicut medical college. The health care workers treating were given full PPE kit and educated regarding personal protection, hand hygiene, waste disposal and hospital sanitization. They were given 14 days duty followed by 14 days quarantine when they were not permitted to mingle with others including family members. All private doctors' clinics and hospitals were asked to report any suspected covid cases to the government run health care system. A few private clinics where COVID-19 patients reported first were closed down as a precautionary measure.

On 10th March, the Kerala government arranged special isolation wards in prisons across the state. On 10th March, the government of Kerala shut down all colleges and schools. Online education was encouraged. The government also urged people to not undertake pilgrimages, attend large gatherings such as weddings and cinema shows. Also, the government has launched a mobile application called GoK Direct for users to get information and updates regarding the COVID-19 disease. It is an initiative from the Kerala Startup Mission and the Information & Public Relations Department. The app can also send text message alerts to basic phones (without internet).

On 15th March, a new initiative 'Break the Chain' was introduced by Government of Kerala. The campaign aims to educate people about the importance of public and personal hygiene. The government has appealed to the public to promote *break the chain* campaign as a safety measure. During this campaign, the government has installed water taps with hand wash bottles at public spots of the railway stations and other public places.

On 23rd March, Chief Minister announced a statewide lockdown till 31st of March to prevent further spread of COVID-19, which was later extended up to May 3rd. This was before the central government declared a nationwide lockdown. Being strictly applied in Kasarkode district which was most affected with COVID-19. Necessary shops like grocery stores were allowed to be opened till 11am till 5pm. And in other districts necessary shops were opened from 7 am till 5 pm with the exception of medical stores which was permitted to work round the clock. Public transports were shut down. During the lockdown period police force was widely used to make sure that no one gets out of the house unless in dare emergency. Several cases were charged against those who were getting out of house without valid reason. Their vehicles were also ceased by the government. Even drones were used to make sure that no public gathering is happening. Cases were charged against temple, church and mosque authorities who were breaking the rule of social isolation. These measures drastically brought down the occurrence of new cases in Kerala. As on 17th March only 7 new cases were reported in Kerala. The discharged cases were advised another 14 days of home quarantine. Community kitchens were started under control of all 14 districts administrations to take care of underprivileged citizens.

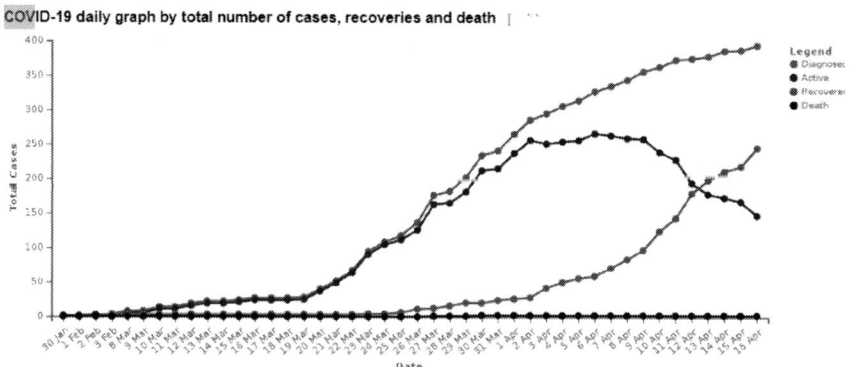

Currently The Union Health Ministry has identified seven districts in Kerala as COVID-19 hotspots. The state is divided into 4 zones red, orange, yellow and green depending on case density. Strict lockdown is being continued in red zone.

REFERENCES

[1] Paul A, Chatterjee S, Bairagi N. Prediction on COVID-19 epidemic for different countries: Focusing on South Asia under various precautionary measures. *medRxiv* 2020.

[2] El Alaoui A. *How Countries of South Mitigate COVID-19: Models of Morocco and Kerala, India.* 2020.

[3] Gupta R, Pal SK. Trend Analysis and Forecasting of COVID-19 outbreak in India. *medRxiv* 2020.

Chapter 11

HEALTH CARE PROFESSIONALS AND COVID-19

INTRODUCTION

Healthcare workers are in forefront in the fight against COVID-19 pandemic. World over, they face several challenges, the risk to their health being the foremost. Health workers have long working hours and are separated from their families for several days. Associated stigma, physical and psychological violence and fatigue add to their psychological stress. Burn out of health care workers need to be prevented. Protecting the health care workers is a priority in such pandemics, which every healthcare system needs to address.

RIGHTS AND RESPONSIBILITIES OF HEALTH CARE WORKER

Health care workers have the right for information about the ongoing pandemic, and the occupational risks involved [1]. They should be made aware of infection control measures, personal protective devices and their proper usage. There should be adequate supply of PPE - masks, gloves, goggles, gowns, hand sanitizer, soap and water, cleaning supplies etc. Also provided appropriate security measures as needed for personal safety. They should be provided a blame free environment to report any occupational exposure they had, proper working hours and adequate rest. They shall be given free treatment in case of infection with COVID-19. Also provided with adequate psychological support if needed with facilities including counseling [1].

PPE AND ITS USE

Health care workers involved in the care of patients with COVID-19 should be using PPE appropriately; this involves selecting proper PPE and being trained in how to put on, remove, and dispose it off. PPE includes gloves, medical masks, goggles or a face shield, and gowns, as well as for specific procedures, respirators (i.e., N95 or FFP2) and aprons. There is shortage of PPE all over the world due to high demand during this pandemic and hence its use must be judicious. In order to minimize the use of PPE, use telemedicine instead of direct interview with the suspected cases of COVID-19. Use physical barriers like glass kiosk where ever possible. Restrict the number of visits to patient with COVID-19. Use of robots for drug and food delivery is a novel useful approach. Type of PPE use will depend on activity involved. Health care workers involved in direct care of patients should wear gown gloves, medical mask, and eye protection (goggles or face shield). When aerosol-generating procedures (e.g., tracheal intubation, non-invasive ventilation, tracheostomy, cardiopulmonary resuscitation, manual ventilation before intubation, bronchoscopy) health care workers should use respirators, eye protection, gloves and gowns;

aprons should also be used if gowns are not fluid resistant. Adhere to PPE donning and doffing protocol [2].

HOME TO WORK

Avoid wearing any accessories, minimize the items you bring for work, bring food in disposable bags consider placing your phone in disposable plastic bag and often disinfect. Use separate shoe for work. Hand hygiene to be followed either using soap and water or alcohol containing sanitizer.

WORK TO HOME

Sanitize hands, and wipe phone, on entering the car clean Steering wheel dash board and other hand touched areas of the car. At home create a decontamination zone (ideally in the garage). Keep a sanitizer, shoe rack, decontamination wipes to clean phone, credit card or any other objects with you. Belts are better avoided as it is difficult to decontaminate them. Proceed directly to shower. Clothes are disinfected with detergent and water above 160 degree fahrenheit. Toilets should be flushed with closed lids.

PRECAUTIONS IN THE CLINICS

Keep the patient seated at least 1 meter distance. If the patient has to be examined use disposable gloves. Avoid examining mouth nose and throat; clean the seat and table with sodium hypochlorate solution. Provide hand sanitizers outside consulting room. Instruct patients to wear masks before entering the consultation room. Avoid crowding in waiting area by giving spaced appointments. Social distancing should be ensured in all stages. Registration counter nurses should wear triple layer mask. Frequently disinfect counter top, mobile phone, door knob, and other frequently touched surfaces.

REFERENCES

[1] World Health Organization. *Coronavirus disease (COVID-19) outbreak: rights, roles and responsibilities of health workers, including key considerations for occupational safety and health, 18 March 2020.* 2020.

[2] World Health Organization. Infection prevention and control during health care when COVID-19 is suspected: interim guidance, 19 March 2020. *World Health Organization*; 2020.

Chapter 12

PSYCHOLOGICAL IMPACT OF COVID-19

INTRODUCTION

The pandemic of COVID-19 has created panic amongst people, health care workers and among health care managers. WHO has created a series of messages that can be used in communications to support mental and Psycho-social well-being in different target groups during the outbreak [1]. Psycho-social support by professionals is often required and should be provided as and when needed for the sufferers.

GENERAL POPULATION

People who are affected by COVID-19 have not done anything wrong, and they deserve our support, compassion and kindness. So do not attach the disease to any particular country or ethnicity. Do not address them as cases; instead use the term "people being treated for COVID-19." In order to avoid stigma on recovery patient should not be addressed as covid patient. People who have undue anxiety about the disorder should refrain from watching news about COVID-19. Seek information only from reliable source. Protect yourself and be supportive to others in the community. Support and honor and appreciate the efforts of health care workers fighting against Covid. There should be a regular mechanisms for surveillance, reporting, and intervention, particularly, when it comes to domestic violence and child abuse [2].

HEALTH CARE WORKERS

Feeling stressed and lonely during such unusual work environment is quite common. Managing the mental health is equally important as managing the physical health [3]. Ensure sufficient rest. Communicate with friends and relatives. Engage in some physical activity; avoid using alcohol, tobacco or other drugs to counter stress. Some healthcare workers may unfortunately experience avoidance by their family or community owing to stigma or fear. This can make an already challenging situation worse. Keep communications alive at least through electronic media.

MANAGERS/TEAM LEADERS

Keep the staff protected from chronic stress. Ensure good quality communication with subordinates. Implement flexible working hours, stress free environment, avoid blames and provide personal protection for employees. Ensure that staff is aware of where and how they can access mental health and psychosocial support services and facilitate access to such services. Required medications like antipsychotics antidepressants must be made available.

CARE GIVERS OF CHILDREN

Address the anxiety of children, as far as possible avoid separation from parents, if child need to be separated suitable alternate to be arranged. Communicate with children if separation is unavoidable by methods like video call. For school going children in lockdown stage, there is a need to implement routines [4], so that they have access to regular programmed work. Online substitutes for daily routines [2], online classes and assignments to be given so as to keep them engaged.

ELDERLY AND DEPENDENT PEOPLE

Elders, especially in isolation and those with cognitive decline/dementia, may become more anxious, angry, stressed, agitated and withdrawn during the outbreak or while in quarantine. They should be provided with practical and emotional support through informal networks (families) and health professionals. They will need help if needed, like calling a taxi, having food delivered and requesting medical care. Make sure that they have up to two weeks of all regular medicines that they may require.

PEOPLE IN ISOLATION

Stay connected via telephone, social media email etc. Encourage them to do regular exercise, and also to read and watch television. Make sure that regular medications are available and taken on time.

REFERENCES

[1] World Health Organization. Mental health and psychosocial considerations during the COVID-19 outbreak, 18 March 2020. *World Health Organization*; 2020.

[2] Galea S, Merchant RM, Lurie N. The Mental Health Consequences of COVID-19 and Physical Distancing: The Need for Prevention and Early Intervention. *JAMA Internal Medicine* 2020.

[3] Zhou X. Psychological crisis interventions in Sichuan Province during the 2019 novel coronavirus outbreak. *Psychiatry Research* 2020;286:112895.

[4] Lee J. Mental health effects of school closures during COVID-19. *The Lancet Child & Adolescent Health* 2020.

Chapter 13

SOCIO - ECONOMIC IMPACT OF COVID-19

INTRODUCTION

The Great Lockdown, following corona virus pandemic is causing an ongoing severe global economic recession which began affecting the world economy in early 2020. The recession is considered to be the steepest economic crisis since the great depression in 2008. On 14th April 2020, the international monitory fund reported that all of the G7 nations had already entered or were entering into what was called a 'deep recession,' alongside most of the western world with significant slowdown of growth across developed and developing economies. The IMF has stated that the economic decline is 'far worse' than that of the great recession in 2008-2009. In order to prevent the spread of COVID-19 more than a third of the world's population at the time is being placed under lockdown. This has caused severe economic repercussions as most of the industries and markets have been shutdown.

As of April 2020, the recession has seen staggering unemployment throughout the world [1, 2] The UN predicts that global unemployment will wipe out 6.7 per cent of working hours globally in the second quarter of 2020 – equivalent to 195 million full-time workers [3].

Unemployment is expected to be at around 10%, with more severely affected nations from the pandemic having higher unemployment rates.

The sudden collapse of the oil price [4] was triggered by the lock down, the collapse of the tourism industry, hospitality and energy and a significant downturn in consumerism in comparison to the previous decade. Stock markets around the world has crashed around 20 to 30% during late February and March 2020, respectively [5]. During the crash, global stock markets made unprecedented and volatile swings, mainly due to extreme uncertainty in the markets.

The aviation industry is severely affected due to the cancellation of the flights, travel restrictions, significant reductions in passenger numbers has resulted in planes flying empty between airports and the cancellation of flights [6]. The cruise ship industry has also been heavily affected by a pandemic, with the share prices of the major cruise lines down 70–80%. Many major sports events including Olympics have been postponed resulting in huge financial loss.

The COVID-19 economic crisis is more akin to a war, wherein governments force the shutdown of significant portions of society with an unknown duration in response to a factor beyond their control. As the underlying problem is not financial, there is no monetary policy response that can fully address it [7]. In addition, the scope of the COVID-19 crisis is, affecting all geographical regions and industries. As a different kind of problem, the COVID-19 crisis will require a totally different type of solution. While creating new opportunities for a few innovative industries to respond to new market demands and flourish. As in times of historical economic crisis, the restructuring process— guided by restructuring advisors and the wisdom and experience of bankruptcy judges—will be necessary to help markets deal with the aftermath of this crisis by providing for economic recovery and a prosperous future [7].

REFERENCES

[1] Gangopadhyaya A, Garrett AB. *Unemployment, Health Insurance, and the COVID-19 Recession*. Health Insurance, and the COVID-19 Recession (April 1, 2020) 2020.

[2] Huang Y, Loungani P, Wang G, Freeman R, Li X. *The Chinese labour market: High unemployment coexisting with a labour shortage*. 2019.

[3] Nelson B, Pettitt AK, Flannery J, Allen N. *Psychological and Epidemiological Predictors of COVID-19 Concern and Health-related Behaviors*. 2020.

[4] Albulescu C. *Coronavirus and oil price crash*. Available at SSRN 3553452 2020.

[5] Toda AA. Susceptible-Infected-Recovered (SIR) Dynamics of COVID-19 and Economic Impact. *arXiv preprint arXiv*:2003 11221 2020.

[6] Hoque A, Shikha FA, Hasanat MW, Arif I, Hamid ABA. The Effect of Coronavirus (COVID-19) in the Tourism Industry in China. *Asian Journal of Multidisciplinary Studies* 2020;3(1).

[7] Rapisardi JJ, Beiswenger JT. The 2020 *Economic Crisis*.

Chapter 14

FUTURE DIRECTIONS

INTRODUCTION

There are several studies going on in different parts of the world to prevent and cure COVID-19. The main 3 avenues that are being used include repurposed drugs and novel drugs, antibodies and vaccines. Several trials have given promising results and several studies are underway and expected to give their results in near future.

DRUG DEVELOPMENT

Repurpose drug is the method where drugs which were approved for other diseases being tested for COVID-19. These include chloroquine, hydroxycholoroqine [1], azithromycin [1], ivermectin, Lopinavir/ritonavir [2], nitric oxide, ascorbic acid, interferon beta 1 b, sarilumab, baricitinib, ruxolitinib, tocilizumab, and many more. In future we expect some promising results to come up [3].

ANTIBODIES AGAINST COVID-19

Transferring antibodies [4] against the virus is another mode to attack the virus. This line of research aims at isolating key antibodies active on the fight against COVID-19, for instance starting from the serum derived from cured patients [5] or from live animals (e.g., mice). The antibodies may then be injected in infected patients to reduce the symptoms of COVID-19. In the future, antibodies may also potentially be injected in individuals at risk in order to reduce risk of infection.

VACCINES

There are several vaccine trials going around the world [6]. The major platform targets advanced into Phase I safety studies include:

- Nucleic acid (DNA and RNA)
- Viral vector
- Virus-like particle involved in DNA replication

During the global emergency of the COVID-19 pandemic, strategies are under consideration to fast-track the timeline for licensing a vaccine against COVID-19, especially by compressing (to a few months) the usually lengthy duration of Phase II-III trials (typically, many years). An effective vaccine will be the best option for controlling this pandemic and its resurgence [7].

REFERENCES

[1] Gautret P, Lagier JC, Parola P, Meddeb L, Mailhe M, Doudier B, et al. Hydroxychloroquine and azithromycin as a treatment of

COVID-19: results of an open-label non-randomized clinical trial. *International journal of antimicrobial agents* 2020;105949.

[2] Cao B, Wang Y, Wen D, Liu W, Wang J, Fan G, et al. A trial of lopinavirGÇôritonavir in adults hospitalized with severe Covid-19. *New England Journal of Medicine* 2020.

[3] World Health Organization. *Coronavirus disease 2019 (COVID-19): situation report*, 72. 2020.

[4] Shanmugaraj B, Siriwattananon K, Wangkanont K, Phoolcharoen W. Perspectives on monoclonal antibody therapy as potential therapeutic intervention for Coronavirus disease-19 (COVID-19). *Asian Pac J Allergy Immunol* 2020;38(1):10-8.

[5] Chen L, Xiong J, Bao L, Shi Y. Convalescent plasma as a potential therapy for COVID-19. *The Lancet Infectious Diseases* 2020;20(4):398-400.

[6] Prompetchara E, Ketloy C, Palaga T. Immune responses in COVID-19 and potential vaccines: Lessons learned from SARS and MERS epidemic. *Asian Pac J Allergy Immunol* 2020;38(1):1-9.

[7] Negahdaripour M. The Battle Against COVID-19: Where Do We Stand Now? *Iranian journal of medical sciences* 2020;45(2):81.

INDEX

2

2019 novel coronavirus, vii, 7, 8, 9, 11, 18, 19, 25, 26, 32, 38, 39, 46, 53, 55, 82, 87, 102
2019-nCoV, vii, 3, 5, 8, 9, 12, 18, 19, 28, 29, 31, 39, 45, 82

A

abdominal discomfort, 4
access, 73, 100, 101
acute respiratory distress, 1, 4, 35, 39
acute respiratory distress syndrome, 1, 4, 35, 39
adults, 60, 69, 73, 74, 75, 77, 81, 109
aerosols, 4, 48, 71
age, 3, 15, 22, 24, 57, 61, 66, 69, 73, 74, 75, 78, 85, 86
air travel, 2, 7
anorexia, 4, 76
antibody, 5, 41, 43, 46, 61
antibody test kits, 5, 41
anxiety, 75, 99, 101

ARDS, 6, 22, 23, 24, 34, 44, 57, 61, 70, 72, 77, 79
assessment, 72, 76, 77
asymptomatic, 3, 4, 5, 6, 12, 16, 24, 26, 41, 47, 59, 85
asymptomatic individuals, 3
authorities, 12, 21, 92

B

back pain, 4
bacterial infection, 6, 58, 67, 69
bilateral, 5, 35, 44, 45, 77
blood, 5, 24, 31, 34, 52, 64, 75, 77, 87
blood pressure, 64, 75, 77
breathing, 22, 48, 66, 67, 68, 69, 70, 76, 77

C

cancer, 57, 64, 86
capillary, 73, 74, 78
capillary refill, 73, 74, 78
cardiac failure, 4, 77
cardiovascular disease, 15, 57, 64
CDC, 1, 7, 19, 46

challenges, 39, 74, 95
chest discomfort, 4
children, 59, 69, 72, 73, 74, 75, 77, 78, 101
China, viii, 1, 2, 7, 8, 9, 11, 12, 14, 15, 16, 17, 18, 19, 25, 26, 38, 46, 83, 86, 87, 105
Chinese Center for Disease Control and Prevention (China CDC), 1, 7, 14, 19, 25, 26, 46
chloroquine, 5, 9, 57, 58, 59, 80, 81, 82, 107
common symptoms, 4, 21, 22
community, 6, 12, 41, 47, 65, 66, 89, 90, 100
complications, 22, 65, 77, 80
contact history, 6, 41
corona, vii, 2, 3, 5, 12, 14, 22, 27, 28, 30, 33, 59, 90, 103
corona care centre, vii
coronavirus, vii, 7, 8, 9, 11, 12, 17, 18, 19, 25, 26, 31, 32, 38, 39, 45, 46, 49, 53, 54, 55, 75, 80, 82, 83, 84, 87, 102
cough, 4, 6, 21, 22, 48, 64, 66, 67, 76, 77
counseling, 75, 90, 96
COVID-19, v, vi, vii, viii, 1, 2, 3, 4, 5, 7, 8, 9, 11, 12, 14, 15, 16, 17, 18, 19, 21, 22, 23, 24, 25, 26, 27, 31, 33, 34, 35, 36, 38, 39, 41, 42, 43, 44, 45, 46, 47, 48, 49, 50, 51, 52, 53, 54, 55, 57, 58, 59, 60, 61, 62, 63, 64, 65, 66, 69, 76, 78, 80, 81, 82, 83, 84, 85, 86, 87, 89, 90, 91, 92, 93, 95, 96, 98, 99, 102, 103, 104, 105, 107, 108, 109
CRP, 62, 68, 76
CT scan, 5, 44, 45, 77
cyanosis, 64, 69, 77

D

death rate, viii, 2, 15, 85, 86, 90
deaths, 15, 89, 90
detection, 3, 12, 41, 50, 53
diabetes, 15, 22, 24, 57, 64, 65, 66, 86
diarhoea, 4
diarrhea, 22, 49, 60, 64
diseases, 2, 24, 32, 53, 59, 83, 86, 107
distress, 1, 64, 69, 70, 71, 77
droplet spread, 3
drugs, 5, 30, 36, 57, 58, 60, 63, 82, 100, 107
dry cough, 4
dyspnea, 22, 76, 86
dyspnoea, 4, 23

E

economic burden, vii
education, 6, 55, 91
elderly patients, 3
emergency, 12, 14, 18, 55, 66, 68, 69, 70, 79, 82, 92, 108
environment, 48, 51, 96, 100
enzyme, 22, 30, 38
epidemic, 1, 9, 12, 14, 16, 17, 33, 58, 81, 93, 109
epidemiology, 7, 14, 17, 32, 38
epithelial cells, 22, 31, 49
equipment, 50, 51, 55
evidence, 12, 23, 24, 36, 41, 49, 58, 62, 70, 75, 77
exposure, 4, 21, 52, 96

F

families, 28, 90, 95, 101
fatigue, 4, 22, 76, 86, 95
FDA, 61, 63, 84
fever, 4, 21, 22, 23, 24, 64, 66, 67, 76, 77
flu like respiratory symptoms, 1
fluid, 6, 57, 70, 73, 74, 75, 77, 97
food, 59, 96, 97, 101
fusion, 29, 30, 31, 34, 72

G

genes, 28, 29, 42
genome, 3, 8, 11, 27, 28, 29, 33, 34, 42
global population, vii
guidance, 9, 14, 43, 45, 46, 53, 55, 84, 98
guidelines, 41, 43, 64, 69, 80

H

hand hygiene, 6, 50, 91, 97
hand washing, 6, 51
headache, 21, 22, 76, 86
health, vii, 2, 12, 14, 15, 17, 18, 21, 42, 46, 51, 59, 89, 90, 95, 96, 98, 99, 100, 101, 102
health care, 14, 42, 46, 51, 59, 89, 90, 95, 96, 98, 99, 100
health care workers, 6, 14, 42, 51, 59, 90, 95, 96, 99, 100
history, 6, 41, 65, 66, 67, 86
HIV, 5, 57, 64, 65
host, 27, 29, 31, 33, 77
Huanan Seafood and wet animal Market, 1
human, viii, 3, 9, 11, 12, 27, 29, 30, 31, 32, 45, 58, 72
hygiene, 6, 50, 91, 97
hypertension, 15, 22, 24, 57, 64, 86
hypoxemia, 60, 70, 71, 73
hypoxia, 23, 24, 76

I

immunosuppression, 24, 26, 39, 83
in vitro, 9, 58, 82, 84
incidence, 12, 17, 22, 35, 36
incubation period, 3, 16, 21, 47
India, vii, viii, ix, 5, 36, 55, 59, 60, 64, 89, 90, 93

individuals, 3, 16, 21, 22, 30, 50, 55, 85, 86, 108
infants, 3, 55, 66, 73, 78
infection, vii, 3, 4, 5, 6, 8, 11, 12, 14, 15, 16, 21, 22, 23, 24, 25, 26, 31, 33, 34, 35, 36, 39, 42, 43, 44, 45, 46, 50, 51, 52, 53, 58, 63, 67, 69, 73, 76, 77, 84, 85, 87, 89, 95, 98, 108
influenza, 6, 9, 60, 64
injury, 36, 39, 49, 52, 72, 86
inter personal transmission, 2
intervention, 6, 100, 109
isolation, 2, 6, 12, 14, 41, 50, 51, 57, 80, 85, 89, 90, 91, 92, 101
Italy, viii, 2, 15, 16, 86, 87

K

kidney(s), 23, 35, 64, 66

L

lead, 42, 48, 68, 74
liver, 4, 23, 34, 35, 36, 49, 64, 66, 76
liver involvement, 4
lopinavir, 5, 60, 107
loss of smell, 4
loss of taste, 4

M

management, 6, 23, 39, 50, 55, 57, 84
masks, 6, 67, 96, 97
mechanical ventilation, 23, 70, 71, 72, 74, 76
media, 7, 19, 100, 101
medical, 12, 13, 14, 46, 50, 52, 55, 66, 85, 91, 92, 96, 101, 109
medicine, 9, 18, 25, 26, 39, 87

mortality, viii, 2, 4, 9, 17, 24, 25, 26, 46, 55, 57, 59, 64, 85, 86, 87, 90
myalgia, 4, 22
myocarditis, 4, 22, 26, 34, 87

N

nausea, 4, 60, 76
New England, 8, 18, 25, 54, 55, 81, 109
nosocomial transmission, 1
novel corona virus, vii, 1, 2, 8, 11, 12, 33
novel corona virus infection, vii

O

oedema, 35, 71, 74, 77
organ(s), 4, 23, 35, 36, 71, 77
oxygen, 23, 61, 62, 63, 67, 68, 69, 70, 71, 76, 77, 78

P

pain, 4, 22, 64, 66, 67, 76
pandemic, vii, 2, 6, 14, 15, 16, 19, 33, 36, 48, 53, 54, 55, 81, 89, 95, 96, 99, 103, 104, 108
pathogenesis, 7, 31, 32, 33, 34, 38
patients with comorbidities, 3
PCR, 41, 42, 43, 46, 79, 90
perfusion, 70, 73, 74, 75
pericarditis, 4
person to person transmission, 3, 57
personal protective device, 6, 96
pneumonia, 1, 4, 7, 8, 9, 13, 17, 18, 22, 23, 24, 25, 32, 34, 39, 46, 58, 60, 66, 76, 77, 81, 82, 83, 86, 87
polymerase, 4, 5, 60, 82
population, vii, 2, 85, 89, 103
prevention, 14, 43, 46, 55, 78, 98
professionals, 50, 57, 99, 101

prophylaxis, 58, 59, 78, 81
protease inhibitors, 5, 57, 60
protection, 50, 57, 79, 91, 96, 100
public health, 2, 12, 14, 89
public health services, 2

R

receptor, 7, 12, 22, 26, 29, 30, 31, 32, 34, 38
receptors, 30, 31, 34, 35
recommendations, iv, 9, 18, 41, 54, 60, 80, 89
recovery, 46, 58, 83, 99, 104
remdesivir, 5, 9, 61, 68, 82, 83
renal failure, 4
replication, 29, 31, 32, 33, 38, 61, 84, 108
requirement, 62, 63, 67, 68, 76
respiratory distress syndrome, 1, 4, 35, 39
respiratory failure, 24, 34, 70, 72, 74, 77
response, 1, 8, 14, 15, 62, 68, 74, 75, 77, 104
risk(s), 14, 25, 46, 48, 58, 65, 66, 70, 72, 77, 85, 95, 108
risk factors, 25, 46, 65
RNA, 3, 27, 28, 32, 34, 41, 48, 53, 60, 82, 108
runny nose, 4

S

safety, 51, 58, 62, 80, 91, 96, 98, 108
SARS, vii, 2, 8, 12, 13, 22, 25, 26, 27, 29, 30, 31, 32, 33, 34, 36, 38, 46, 47, 49, 53, 54, 55, 58, 60, 82, 84, 86, 87, 109
SARS-CoV, vii, 2, 8, 12, 13, 22, 25, 27, 29, 30, 31, 32, 34, 36, 46, 49, 53, 54, 55, 58, 60, 82, 84, 86, 87
SARS-CoV-2, vii, 2, 8, 12, 13, 22, 25, 27, 32, 34, 36, 46, 54, 55, 58, 60, 82, 84, 86, 87
saturation, 65, 67, 77

school, 14, 91, 101, 102
septic shock, 23, 73, 74
serology, 28, 43, 46
serum, 43, 76, 78, 108
services, 2, 17, 90, 100
severe acute respiratory syndrome corona virus 2, vii, 2, 27
shock, 69, 70, 73, 74, 75
signs, 64, 65, 66, 67, 69, 72, 73, 74, 75, 76, 77
skin, 52, 73, 74, 77, 78
sneezing, 4, 50
social distancing, 6, 16, 85, 89, 98
social isolation, 2, 6, 50, 89, 92
solution, 52, 97, 104
sore throat, 4, 22, 64, 76
sputum, 4, 22, 42, 76, 86
state, 35, 61, 64, 89, 90, 91, 92
steroids, 58, 62, 63, 64
stigma, 95, 99, 100
stroke, 22, 36, 74, 86
symptoms, 1, 4, 14, 21, 22, 23, 24, 47, 48, 49, 65, 66, 69, 76, 77, 79, 90, 108
syndrome, 1, 4, 23, 24, 27, 35, 39, 77, 87

T

tachyarrythmeas, 4
tachycardia, 4, 65, 73, 78
tachypnea/tachypnoea, 4, 73, 78
target, 31, 42, 53, 69, 70, 71, 75, 99
temperature, 49, 65, 77, 79
testing, 4, 9, 43, 45, 82, 89, 90
therapy, 67, 69, 70, 71, 78, 83, 109
thrombosis, 23, 36, 78
tissue, 34, 35, 36, 50, 75
transcription, 4, 29, 42
transmission, 1, 2, 3, 5, 6, 11, 12, 16, 17, 18, 24, 25, 27, 29, 47, 48, 49, 54, 55, 57, 65, 66, 71

travel ban, 2
treatment, 6, 9, 12, 14, 25, 30, 36, 39, 48, 55, 57, 58, 60, 66, 67, 69, 80, 81, 82, 83, 84, 96, 108
trial, 5, 9, 59, 61, 62, 71, 81, 82, 109

U

upper respiratory tract, 34, 42, 76
urine, 42, 66, 74, 77

V

vaccine, 5, 17, 53, 57, 85, 108
ventilation, 4, 23, 48, 50, 57, 58, 70, 71, 72, 73, 74, 76, 79, 96
viral infection, 31, 76, 80
viruses, 2, 3, 22, 28, 33

W

war, viii, 90, 91, 104
water, 6, 50, 52, 91, 96, 97
wear, 50, 96, 97
WHO, vii, 1, 2, 4, 7, 9, 12, 14, 19, 41, 43, 50, 53, 61, 76, 89, 99
workers, 6, 14, 42, 51, 52, 59, 90, 95, 96, 98, 99, 100, 103
working hours, 95, 96, 100, 103
World Health Organization (WHO), vii, 1, 2, 4, 7, 9, 12, 14, 18, 19, 27, 32, 41, 43, 45, 46, 50, 53, 54, 55, 61, 76, 82, 84, 89, 98, 99, 102, 109
Wuhan, vii, 1, 2, 7, 8, 9, 11, 12, 13, 14, 17, 18, 25, 26, 27, 32, 38, 46, 58, 83, 87, 89
Wuhan city, 1, 11, 14
Wuhan epidemic, 1